Alexander A. Friedmann

Papers On Curved Spaces and Cosmology

Translated and edited by Vesselin Petkov

MINKOWSKI
Institute Press

Alexander Friedmann
29 June 1888 – 16 September 1925

Cover: Based on a photograph from:
http://publ.lib.ru/ARCHIVES/F/FRIDMAN_Aleksandr_Aleksandrovich/

© Minkowski Institute Press 2014
All rights reserved. Published 2014

ISBN: 978-1-927763-22-3 (softcover)
ISBN: 978-1-927763-23-0 (ebook)

Minkowski Institute Press
Montreal, Quebec, Canada
http://minkowskiinstitute.org/mip/

For information on all Minkowski Institute Press publications visit our website at http://minkowskiinstitute.org/mip/books/

Preface

This volume contains three papers by the famous Russian physicist Alexander A. Friedmann on curved spaces and cosmology.

The first two papers "On the Curvature of Space" and "On the Possibility of a World with a Constant Negative Curvature of Space" were first published in German in *Zeitschrift für Physik* in 1922[1] and 1924,[2] respectively. The first paper was published in Russian in *Журнал Русского физико-химического общества* (Zhurnal Russkogo Fiziko-Chimicheskogo Obshtestva – Journal of the Russian Physico-Chemical Association) in 1924[3], and reprinted, together with the second paper, in *Успехи физических наук* (Uspekhi Fizicheskikh Nauk – Achievements of the Physical Sciences) in 1963[4]

The third paper "On the Geometry of Curved Spaces" is a 28-page typed manuscript (dated 15 April 1922) which has not been published even in Russian. It has been preserved, together with the original typed manuscript (in Russian) of Friedmann's 1922 paper "On the curvature of space" and several letters by Friedman and one by Jan Schouten, in the Ehrenfest archive[5] (http://www.lorentz.

---

[1] A. Friedman, Über die Krümmung des Raumes, *Zeitschrift für Physik* **10** Nr. 1, 1922, S. 377-386. In this first paper Friedmann's name is not properly transcribed; he himself wrote his name in Latin alphabet with two 'n'.

[2] A. Friedmann, Über die Möglichkeit einer Welt mit konstanter negativer Krümmung des Raumes, *Zeitschrift für Physik* **21** Nr. 1, 1924, S. 326-332.

[3] А. А. Фридман, О кривизне мира, *Журн. Русск. физ.-химич. об-ва*, Ч. физ., 1924/25 т. 56 вып. 1 стр. 59-68

[4] А. А. Фридман, О кривизне мира, *УФН* (1963) **80**, вып. 3, стр. 439-446;

А. А. Фридман, О возможности мира с постоянной отрицательной кривизной пространства, *УФН* (1963) **80**, вып. 3, стр. 447-452.

[5] The text introducing Friedmann's papers in the archive reads: "Alexander Friedmann (1888-1925) has been called the man who made the universe expand, because he first raised the possibility of the expansion of the universe (discovered later by Hubble). Friedmann was a friend of Paul Ehrenfest, from their years together in St. Petersburg. The Ehrenfest archive at the Instituut-Lorentz contains several letters and manuscripts that Friedmann sent to Ehrenfest in 1922. (This is the same archive in which the Einstein manuscripts were found.) The seminal 1922 paper is here in a Russian original, as well as an unpublished manuscript on the geometry of curved space. This material is presented here for the first time.

leidenuniv.nl/history/Friedmann_archive/). Friedmann outlines the ideas in the unpublished manuscript and his reaseach in a letter sent to Ehrenfest on 15 June 1922. This letter is among the letters preserved in the Ehrenfest archive (http://www.lorentz.leidenuniv.nl/history/Friedmann_archive/Friedmann_Ehrenfest/Friedmann_to_Ehrenfest.pdf).

There exist two English translations of Friedmann's 1922 and 1924 papers – in *Cosmological Constants: Papers in Modern Cosmology*[6] in 1986 and in the journal *General Relativity and Gravitation*[7] in 1999. These two English translations were done from the German publications and when compared to the original Russian text it is evident that, due to the double translation (from Russian to German, and from German to English), (i) at some places the meaning of Friedmann's explanations is not properly conveyed and (ii) often leave the impression that Friedmann did not express himself clearly, whereas the original Russian text is clear. For this reason, these two papers were now translated directly from the original Russian texts.

The paper "On the Curvature of Space" and the unpublished manuscript "On the Geometry of Curved Spaces" were translated from the typed Russian manuscripts kept in the Ehrenfest archive. The translation of the unpublished manuscript turned out to be significantly more difficult because at many places in this paper Friedmann used long sentences which are difficult to read even in Russian (like long sentences in any language). An English translation by Dmitry Pikulin, which has been recently added to the Ehrenfest archive, and which was consulted, could not help much.

The direct translation from the Russian of Friedmann's 1922 and 1924 papers provides sufficient justification for the publication of their

---

The originals are kept in the archives of the Museum Boerhaave in Leiden."

[6] J. Bernstein and G. Feinberg (eds.), *Cosmological Constants: Papers in Modern Cosmology* (Columbia University Press, New York 1986) pp. 49-65.

[7] A. A. Friedmann, On the curvature of space, *General Relativity and Gravitation* **31** (12) (1999) pp. 1991-2000;

A. A. Friedmann, On the Possibility of a World with Constant Negative Curvature of Space, *General Relativity and Gravitation* **31** (12) (1999) pp. 2001-2008.

See also A. Krasiński and G. F. R. Ellis, Editor's Note (on the classic Friedmann papers), *General Relativity and Gravitation* **31** (1999) pp. 1985-1989, and A. Krasiński, Addendum, *General Relativity and Gravitation* **32** (2000) pp. 1937-1938.

new English translations (along with his unpublished manuscript). An additional reason is the one pointed out by A. Krasiński and G. F. R. Ellis in their 1999 Editor's Note to the English translation of Friedmann's two papers in the "Golden Oldies" series of the journal *General Relativity and Gravitation*:[8]

> There is probably not a single relativist who does not know about these two papers, the first published papers on the idea of an expanding and evolving universe. Nevertheless, the number of those who know what the papers actually contain or do not contain is much smaller. Legends have established themselves about what part of the credit belongs to Alexander Friedmann, what part belongs to Georges Lemaître, etc., and false information is being multiplied through citations. Hence, the main purpose of reprinting these not-too-easily accessible papers is to make them more generally available to contemporary physicists and to put some of those persistent legends aside. Also, Friedmann's own presentation has pedagogical value and may be of use in modern courses on relativity.

Friedmann's papers can be made even further available through this small and affordable book, which can be more easily accessible than an issue of a specialized journal (excluding the electronic versions of the papers to which academics have immediate access) and the not-readily-available 1986 collection of papers containing the first English translation of Friedmann's two papers (which in any way turned out to be largely unnoticed by physicists; even the editors of the "Golden Oldies" series of *General Relativity and Gravitation* were unaware[9] of its existence when they published the English translation of Friedmann's papers in 1999).

Montreal, 28 January 2014 *Vesselin Petkov*

---

[8] A. Krasiński and G. F. R. Ellis, Editor's Note (on the classic Friedmann papers), *General Relativity and Gravitation* **31** (1999) pp. 1985-1989.

[9] A. Krasiński, Addendum, *General Relativity and Gravitation* **32** (2000) pp. 1937-1938.

Contents

Preface	i
Introduction	1
On the Curvature of Space	11
On the Possibility of a World with a Constant Negative Curvature of Space	23
On the Geometry of Curved Spaces	31
Appendix — Main dates in Friedmann's life and work	59
Image Credits	63

INTRODUCTION

If the talented Russian physicist Alexander Alexandrovich Friedmann had to be introduced with a single sentence, the most appropriate sentence would be the title of his biography translated from the Russian: *Alexander A. Friedmann: The Man who Made the Universe Expand*.[10]

Indeed, he was the first to realize in 1922 that Einstein's equations have solutions which describe not only a stationary Universe as Einstein initially believed, but also a non-stationary world. Friedmann won the debate with Einstein over the admissibility of such solutions, but his life was too short and he could not see the triumph of his views when the experimental evidence fully supported his predictions and demonstrated that the Universe was expanding.

Friedmann was born on June 4 (16), 1888 in St. Petersburg.[11] His parents had very little to do with science. His father was a ballet dancer, musician and composer, and his mother (the daughter of a famous musics professor in the St. Petersburg Conservatory, composer and conductor) gave piano lessons. Friedmann admitted he did not inherit any musical talent from his parents.

From 1897 to 1906 Friedmann attended the Second St. Petersburg

---

[10]E. A. Tropp, V. Ya. Frenkel, A. D. Chernin, *Alexander A. Friedmann: The Man who Made the Universe Expand*. Translated by A. Dron and M. Burov (Cambridge University Press, Cambridge 1993).

[11]This summary of Friedmann's biography is based on P. Ya. Polubarinova-Kochina, In Memory of A. A. Fridman (On the seventy-fifth anniversary of his birth), *Usp. Fiz. Nauk* **80** (1963) pp. 345-352, V. Ya. Frenkel, Alexander Alexandrovich Friedmann (Biographical sketch) *Usp. Fiz. Nauk* **155** (1988) pp. 481-516, and E. A. Tropp, V. Ya. Frenkel, A. D. Chernin, *Alexander A. Friedmann: The Man who Made the Universe Expand*. Translated by A. Dron and M. Burov (Cambridge University Press, Cambridge 1993). A list of the main dates in Friedmann's life and work is given in the Appendix.

Gymnasium where he impressed his mathematics teachers with his mathematical talent. In 1905, while still at the school, Friedmann and his friend Ya. D. Tamarkin wrote their first scientific paper on the Bernoulli numbers, which was published in 1906.

Not only was Friedmann scientifically active, but he was also involved in the political life of that time. In 1905-1906 he took part in the students' movement of the St. Petersburg secondary schools and was a member of the Central Committee of the student organization.

From 1906 to 1910 Friedmann was a student at the Faculty of Physics and Mathematics of St. Petersburg University. After graduating from the University in 1909, he was invited by two of his professors – V. A. Steklov and D. K. Bobylev – to pursue advanced studies with the possibility to be retained at the University as a professor in the Department of Pure and Applied Mathematics.

Friedmann (first row, second from left) with colleagues and friends. His wife Ekaterina is first (from left) in the second row.

In 1911 Friedmann married Ekaterina Petrovna Dorofeyeva which got by surprise his friends, colleagues and professors. On 13 July 1911 his friend Tamarkin found it nevessary to inform their professor Steklov:[12]

---

[12] E. A. Tropp, V. Ya. Frenkel, A. D. Chernin, *Alexander A. Friedmann: The Man who Made the Universe Expand.* Translated by A. Dron and M. Burov (Cambridge University Press, Cambridge 1993), p. 55.

> The marriage of Alexander Alexandrovich was as unexpected to me as it was to you. His wife is quite a good-looking woman, although slightly older than he is. So far, I can say that the marriage has had only a positive effect on Alexander; it has reduced his habitual nervousness, made him calmer, and in no way hampered our studies, which have continued without interruption five times a week.

Friedmann and his wife often visited Steklov's home which allowed Steklov to note that Ekaterina was a very quiet woman who had a considerable and beneficial effect on her husband. Later, in 1925, "Steklov applied for a pension for Ekaterina Friedmann and highly commended the assistance that she had given to Alexander Friedmann, working on translations of his articles, reading proofs, etc."[13]

In 1914 Friedmann is sent to Leipzig in Germany to do scientific research work under V. Bjerknes.

From 1914 to 1916 Friedmann took part in World War I. He volunteered to join the Army and served in aviation units of the northern and southern fronts.

Even during the war Friedmann did not stop doing research – as a pilot he worked on the theory of precision bombing, compiling the appropriate tables, the use of which had risen dramatically the probability for the bombs to hit the intended target. In a letter to Steklov (dated 28 February 1915) Friedmann discussed his work on the theory of precision bombing:[14]

> I have recently had a chance to verify my ideas during a flight over Przemyśl; the bombs turned out to be falling almost the way the theory predicts. To have conclusive proof of the theory I'm going to fly again in a few days. The bombs I drop (5 lb, 25 lb and 1 pood [40 lb] in weight)

---
[13] *Ibid.*
[14] *Ibid*, p. 72.

belong to the class in which α is very small, so I've been verifying the expansion of the solutions in terms of the parameter α.

After the war Friedman held a number of positions, including Professor and Chair of the Department of Mechanics at the Perm State University (1918-20), Professor in the Faculty of Physics and Mathematics of the Petrograd Polytechnical Institute (1920-25) and from February 1925 – Director of the Main Physical (later Geophysical) Observatory.

Friedmann in his office in the Main Physical Observatory.

In 1923 had enormously difficult time (in his own words) when he divorced his wife Ekaterina and married Natalia Yevgenievna Malinina (1893-1981) – a physicist at the Main Physical Observatory who worked at the time on Earth's magnetism and, particularly, the Kursk magnetic anomaly.

In his scientific career Friedmann worked mostly in the fields of hydrodynamics and metereology. In the

Friedmann with his second wife Natalia Malinina (left)

last several years of his life he directed his intellectual power toward the theory of relativity and its implications for cosmology. As indicated in Friedmann's Biography he and his colleague V. K Frederiks had been intensely working on the theory of relativity[15]:

> From the 1920s there was a regular seminar at the Physical Institute of Petrograd University, where Friedmann and Frederiks presented papers on general relativity.

---

[15]E. A. Tropp, V. Ya. Frenkel, A. D. Chernin, *Alexander A. Friedmann: The Man who Made the Universe Expand*. Translated by A. Dron and M. Burov (Cambridge University Press, Cambridge 1993) p. 115.

On 15 April 1922 Friedmann completed his manuscript "On the Geometry of Curved Spaces" and sent it to Ehrenfest. A bit later (on 29 May 1922) he completed his paper "On the curvature of space", which appeared in *Zeitschrift für Physik* the same year.[16]

In 1923 Friedmann published a book *The World as Space and Time* on the emerging (after the works of Einstein and Minkowski) spacetime view of the world, which was written for a wider audience.[17]

In November 1923 Friedman completed his paper "On the Possibility of a World with a Constant Negative Curvature of Space", which appeared in *Zeitschrift für Physik* the following year.[18]

The cover of Friedmann's book *The World as Space and Time*.

In 1924 Friedmann and Frederiks began to write a fundamental monograph on the theory of relativity[19]:

> They set themselves the task of presenting the theory with adequate rigor from the logical point of view, assuming the reader's background in mathematics and theoretical physics did not exceed the level of knowledge given by Russian universities and higher technical educational institutions. It was originally intended to publish the whole book at once, but technical obstacles made the authors divide the book into five parts and prepare these parts as separate issues. The first issue of the book expounded

---

[16] A. Friedman, Über die Krümmung des Raumes, *Zeitschrift für Physik* **10** Nr. 1, 1922, S. 377-386.

[17] The first English translation of this book will appear in March 2014. See Alexander A. Friedmann, *The World as Space and Time* (Minkowski Institute Press, Montreal 2014) at http://www.minkowskiinstitute.org/mip/books/friedmann1.html

[18] A. Friedmann, Über die Möglichkeit einer Welt mit konstanter negativer Krümmung des Raumes, *Zeitschrift für Physik* **21** Nr. 1, 1924, S. 326-332.

[19] E. A. Tropp, V. Ya. Frenkel, A. D. Chernin, *Alexander A. Friedmann: The Man who Made the Universe Expand*. Translated by A. Dron and M. Burov (Cambridge University Press, Cambridge 1993) p. 117.

the fundamentals of tensor calculus. The second issue was to be devoted to the fundamentals of multi-dimensional geometry, the third to electrodynamics, and, finally, the fourth and fifth to the fundamentals of special and general relativity.

Unfortunately, only the first volume of the monograph was published:[20] V. K. Frederiks and A. A. Friedmann, *Foundations of the Theory of Relativity, Volume 1: Tensor Calculus* (Academia, Leningrad 1924).

The first reaction to Friedmann's work came from Einstein himself and it was negative. In a brief note[21] which was published in *Zeitschrift für Physik* in 1922 he wrote:[22]

The first and only published volume of *Foundations of the Theory of Relativity* by Friedmann and Frederiks.

The work cited contains a result concerning a non-stationary world which seems suspect to me. Indeed, those solutions do not appear compatible with the field equations (A). From the field equations in follows necessarily that the divergence of the matter tensor T_{ik} vanishes. This along with the anzatzes (C) and (D) leads to the condition

$$\frac{\partial \rho}{\partial x_4} = 0$$

---

[20] The first English translation of this book will appear in April 2014. See V. K. Frederiks and A. A. Friedmann, *Foundations of the Theory of Relativity, Volume 1: Tensor Calculus* (Minkowski Institute Press, Montreal 2014) at http://www.minkowskiinstitute.org/mip/books/friedmann3.html

[21] A. Einstein, Bemerkung zu der Arbeit von A. Friedmann Über die Krümmung des Raumes, *Zeitschrift für Physik* **11** (1922), p. 326.

[22] Quoted from: J. Bernstein and G. Feinberg (eds.), *Cosmological Constants: Papers in Modern Cosmology* (Columbia University Press, New York 1986) p. 66.

which together with (8) implies that the world-radius R is constant in time. The significance of the work therefore is to demonstrate this constancy.

On 6 December 1922 Friedmann sent a detailed letter to Einstein. Here is part of it:[23]

> Dear Professor,
>
> From the letter of a friend of mine who is now abroad I had the honor to learn that you had submitted a short note to be printed in the 11th volume of the *Zeitschrift für Physik*, where it is stated that if one accepts the assumptions made in my article "On the curvature of space," it will follow from the world equations derived by you that the radius of curvature of the world is a quantity independent of time.
>
> ...
>
> Should you find the calculations presented in my letter correct, please be so kind as to inform the editors of the *Zeitschrift für Physik* about it; perhaps in this case you will publish a correction to your statement or provide an opportunity for a portion of this letter to be printed.

Friedmann's "a friend of mine" was Yu. A. Krutkov. In May 1923 Krutkov met Einstein at Ehrenfest's home in Leiden and discussed Friedmann's letter with him. In particular, Krutkov drew Einstein's attention to that part of the letter where Friedmann showed by direct calculations that the necessary condition for the disappearance of the divergence of the matter tensor, which was pointed out by Einstein in his note, by no means implies that the radius of the curvature of the world is constant in time. Both Friedmann's letter and Krutkov's explanation convinced Einstein that the results Friedmann's paper were correct and on 21 May 1923 he sent a second note[24] concerning Friedmann's paper to *Zeitschrift für Physik*:[25]

---

[23] E. A. Tropp, V. Ya. Frenkel, A. D. Chernin, *Alexander A. Friedmann: The Man who Made the Universe Expand*. Translated by A. Dron and M. Burov (Cambridge University Press, Cambridge 1993) p. 170.

[24] A. Einstein, Notiz zu der Arbeit von A. Friedmann "Über die Krümmung des Raumes," *Zeitschrift für Physik* **16** (1923), p. 228.

[25] Quoted from: E. A. Tropp, V. Ya. Frenkel, A. D. Chernin, *Alexander A.*

In my previous note I criticized the above-mentioned work. However, my criticism, as I became convinced by Friedmann's letter communicated to me by Mr. Krutkov, was based on an error in calculation. I consider that Mr. Friedmann's results are correct and shed new light. It has turned out that the field equations allow not only static but also dynamic (i.e. variable with respect to time) centro-symmetrical solutions for the space structure.

Friedmann in 1922 or 1923.

Despite that Einstein admitted that his equations have solutions for non-stationary worlds, he did not believe that such solutions reflect something real. During the fifth Solvay meeting at Brussels in 1927 Einstein told Lemaître that in his opinion non-stationary world models are "simply disgusting."[26]

After the successful outcome of the debate with Einstein, the future looked bright and promising for Friedmann. Unfortunately, he had two more years to live. One of the last events he took part was in July 1925 when he participated in a balloon flight which reaching the record elevation of 7,400 m. After that Fiedmann and his wife went on vacation to Crimea. After he returned to Leningrad he became ill. It turned out that when returning from Crimea in August 1925 Friedmann bought and ate pears without washing them and contracted a deadly disease – typhus:[27]

Weakened by the deprivations of the war years and his

Friedmann: The Man who Made the Universe Expand. Translated by A. Dron and M. Burov (Cambridge University Press, Cambridge 1993) p. 172.

[26]T. Jung, Three cosmological dogmas – Einsteins influence on early relativistic cosmology, *Astron. Nachr.* / AN 326 (2005), No. 7, p. 1

[27]E. A. Tropp, V. Ya. Frenkel, A. D. Chernin, *Alexander A. Friedmann: The Man who Made the Universe Expand.* Translated by A. Dron and M. Burov (Cambridge University Press, Cambridge 1993) p. 211.

exhausting work, he failed to throw off the disease, and on September 16, Alexander Alexandrovich Friedmann died.

Friedmann did not live to see the birth of his son:[28]

> Friedmann's son, the third Alexander Alexandrovich in their family, used to say to his mother when he was a teenager: "I want to be an ordinary person." He entered university during the war, served in the Army as a driver, and worked as a driver in Leningrad after the war, until his death in 1983. He had no children.

Let me finish with a final quote from Fridmann's Biography:[29]

> Ekaterina Petrovna Dorofeyeva-Friedmann writes in her memoirs: "Always ready to learn from everybody who knew more than he did, he realized that in his work he was blazing new trials, difficult and unexplored by anyone, and he liked to quote these words of Dante's: 'L'acqua ch'io prendo giammai non si corse' (*Paradiso*, II, 7). 'The sea I sail was never crossed before'."

> In the pages of his research, in the reminiscences of his contemporaries, Friedmann is seen as a profound, independent-minded, and daring thinker who destroys scientific prejudices, myths and dogmas; his intellect sees what others do not see, and will not see what others believe to be obvious but for which there are no grounds in reality. He rejects the centuries-old tradition which chose, prior to any experience, to consider the Universe eternal and eternally immutable. He accomplishes a genuine revolution in science. As Copernicus made the Earth go round the Sun, so Friedmann made the Universe expand.

---

[28] *Ibid*, p. 209
[29] *Ibid*, p. 175

ON THE CURVATURE OF SPACE

§1

1. In their known works on general cosmological questions, Einstein[1] and de Sitter[2] arrive at two possible types of the universe. Einstein obtains the so-called *cylindrical world*, in which space[3] has constant curvature, which does not change with time, wherein the radius of the curvature is associated with the total mass of matter located in space. de Sitter obtains a *spherical world*, in which not only space, but the whole world possesses, to a certain extent, a character of a world of constant curvature.[4] Both Einstein and de Sitter assume a certain form of the matter tensor, reflecting the hypothesis of disconnectedness of matter and its relative rest, in other words, the hypothesis that the velocity of matter is sufficiently small as compared with the fundamental velocity,[5] i.e., with the velocity of light.

The purpose of the present Note is to obtain the cylindrical and spherical worlds as special cases, following from certain general assumptions, and then point out the possibility of obtaining a special world, the curvature of whose space is constant with respect to three

---

[1] A. Einstein, Kosmologische Betrachtungen zur allgemeinen Relativitätstheorie, *Preussische Akademie der Wissenschaften, Sitzungsberichte*, 1917, 142-152.

[2] W. de Sitter, On Einstein's theory of gravitation and its astronomical consequences, *Monthly Notices Roy. Astron. Soc.*, 1916-1917.

[3] By space we will mean a space, described by a manifold of three dimensions, and will assign the term "world" to a space, described by a manifold of four dimensions.

[4] F. Klein, Ueber die Integralform der Erhaltung ersätze und die Theorie der räumlichgeschlossen Welt, *Göttinger Nach.*, 1918.

[5] See this term in Eddington's book: *Espace, Temps et Gravitation*, 2ème partie, Paris, 1921 p. 10.

coordinates, regarded as the space coordinates, but changes with time, i.e. depends on the fourth coordinate, regarded as the time coordinate. As far as its other properties are concerned, this new type of universe resembles Einstein's cylindrical world.

2. Our considerations are based on assumptions grouped into two classes. The first class includes assumptions, identical to those made by Einstein and de Sitter, which are related to the equations governing the gravitational potentials, and to the nature of the state and the motion of matter in space. The second class contains assumptions on the general, so to speak, geometrical nature of our world; both Einstein's cylindrical world and de Sitter's spherical world can be obtained from our hypothesis as special cases.

The assumptions of the first class are the following:

1) The gravitational potentials obey Einstein's equations with the so-called "cosmological" term, which may, in particular, be zero:

$$R_{ik} - \tfrac{1}{2} g_{ik} R + \lambda g_{ik} = -\kappa T_{ik} \quad (i, k = 1, 2, 3, 4), \qquad (A)$$

where g_{ik} are the gravitational potentials, T_{ik} is the matter tensor, κ is a constant, $R = g^{ik} R_{ik}$, and the tensor R_{ik} is defined by

$$R_{ik} = \frac{\partial^2 \ln \sqrt{g}}{\partial x_i \partial x_k} - \frac{\partial \ln \sqrt{g}}{\partial x_\sigma} \left\{ {ik \atop \sigma} \right\} - \frac{\partial}{\partial x_\sigma} \left\{ {ik \atop \sigma} \right\} + \left\{ {i\alpha \atop \sigma} \right\} \left\{ {k\sigma \atop \alpha} \right\}, \qquad (B)$$

where x_i $(i, k = 1, 2, 3, 4)$ are world coordinates, and $\left\{ {ik \atop \sigma} \right\}$ is the Christoffel symbol of the second kind.[6]

2) Matter is in a disconnected state and is at relative rest; or, speaking less rigorously, the relative velocities of matter are negligible compared with the velocity of light. As a result of these assumptions the matter tensor T_{ik} is defined by

$$\begin{aligned} T_{ik} &= 0, \text{ if } i \text{ and } k \text{ are not simultaneously } = 4, \\ T_{44} &= c^2 \rho\, g_{44}, \end{aligned} \qquad (C)$$

where ρ is the density of matter and c is the fundamental velocity; here, of course, the world coordinates are divided into two groups: x_1, x_2, x_3 called space coordinates and x_4 – the time coordinate.

---

[6] Here the sign of R_{ik} and of the scalar curvature R is different compared to the usual sign of these quantities.

3. The assumptions of the second class amount to the following:

1) After separating the three space coordinates (x_1, x_2, x_3) from the four world coordinates, we will have a space of constant curvature, which can, however, change with the fourth time coordinate x_4. The interval ds, defined by $ds^2 = g_{ik} dx_i dx_k$, can be written in the following form with the help of a corresponding change of the space coordinates:

$$ds^2 = R^2(dx_1^2 + \sin^2 x_1 dx_2^2 + \sin^2 x_1 \sin^2 x_2 dx_3^2) + 2g_{14}dx_1 dx_4$$
$$+ 2g_{24}dx_2 dx_4 + 2g_{34}dx_3 dx_4 + g_{44}dx_4^2$$

where R is a function only of x_4; R is proportional to the radius of the curvature of space and therefore the radius of the curvature of space may change with time.

2) In the expression for the interval ds^2, g_{14}, g_{24}, g_{34} vanish if the time coordinate is suitably chosen, or, expressed shortly, time is orthogonal to space. It seems to me that this second assumption does not have any physical or philosophical significance and is introduced solely to simplify the calculations. It is necessary to notice that the worlds of Einstein and de Sitter are special cases of our assumptions.

The assumptions 1) and 2) enable us to write ds^2 in the form

$$ds^2 = R^2(dx_1^2 + \sin^2 x_1 dx_2^2 + \sin^2 x_1 \sin^2 x_2 dx_3^2) + M^2 dx_4^2, \quad (D)$$

where R depends only on x_4, and M is, generally speaking, a function of all four world coordinates. The universe of Einstein is a special case obtained from (D) by replacing R^2 with $-R^2/c^2$ and M with 1, where R is the constant (independent of x_4 as well!) radius of the curvature of space. The universe of de Sitter is obtained when in (D) R^2 is replaced with $-R^2/c^2$ and M with $\cos x_1$:[8]

$$d\tau^2 = -\frac{R^2}{c^2}(dx_1^2 + \sin^2 x_1 dx_2^2 + \sin^2 x_1 \sin^2 x_2 dx_3^2) + dx_4^2, \quad (D_1)$$

---

[7] See, for example, A.S. Eddington, *Espace, Temps et Gravitation*, 2ème partie, Paris, 1921.

[8] Assigning the dimension of time to the interval ds, we denote it by $d\tau$; in this case the constant κ will have the dimension of length divided by mass and in CGS units will be equal to 1.87×10^{-27}. See M. Laue, *Die Relativitätstheorie*, Bd. II, Braunschweig 1921, S. 185.

$$d\tau^2 = -\frac{R^2}{c^2}(dx_1^2 + \sin^2 x_1 dx_2^2 + \sin^2 x_1 \sin^2 x_2 dx_3^2) + \cos^2 x_1 dx_4^2. \quad (D_2)$$

4. It is necessary to say a bit more on the intervals in which the world coordinates are confined; or, in other words, it is necessary to agree on which points of the manifold of four dimensions will be regarded as different. Without going into details, we will assume that the space coordinates change in the intervals: x_1 in the interval $(0, \pi)$; x_2 in the interval $(0, \pi)$; x_3 in the interval $(0, 2\pi)$. As far as the time coordinate is concerned, we will leave the question of the interval in which it changes open and will return to it below.

§2

1. Using equations (A) and (C), assuming that the gravitational potentials are defined by (D), and setting in $i = 1, 2, 3$ and $k = 4$ in equations (A), we find

$$R'(x_4)\frac{\partial M}{\partial x_1} = R'(x_4)\frac{\partial M}{\partial x_2} = R'(x_4)\frac{\partial M}{\partial x_3} = 0.$$

These equations lead to two cases: I) $R'(x_4) = 0$, R does not depend on x_4 and is a constant; we will call the world corresponding to this case a *stationary world*. II) $R'(x_4)$ is not zero, R depends only on x_4; we will call the world corresponding to this second case a *non-stationary world*.

Starting with the stationary world, we write down equations (A) for $i, k = 1, 2, 3$ (assuming $i \neq k$) and obtain the following system of equations:

$$\frac{\partial^2 M}{\partial x_1 \partial x_2} - \cot x_1 \frac{\partial M}{\partial x_2} = 0,$$

$$\frac{\partial^2 M}{\partial x_1 \partial x_3} - \cot x_1 \frac{\partial M}{\partial x_3} = 0,$$

$$\frac{\partial^2 M}{\partial x_2 \partial x_3} - \cot x_2 \frac{\partial M}{\partial x_3} = 0,$$

whose integration gives:

$$M = A(x_3, x_4) \sin x_1 \sin x_2 + B(x_2, x_4) \sin x_1 + C(x_1, x_4), \quad (1)$$

where A, B, C are arbitrary functions of their arguments. Solving equations (A) for the tensor R_{ik} and eliminating the unknown density ρ' from the obtained and still unused equations, and substituting the expression (1) for M in these equations, we find, after somewhat lengthy, but quite elementary calculations, that for M the following two expressions are possible:

$$M = M_0 = \text{const.} \qquad (2)$$

$$M = (A_0 \, x_4 + B_0) \cos x_1, \qquad (3)$$

where M_0, A_0, B_0 are constants.

In the case when M is constant, we have for the stationary world the case of the cylindrical world. In this case, it is more convenient to work with the gravitational potentials of (D$_1$); by determining the density and the quantity λ we obtain Einstein's result:

$$\lambda = \frac{c^2}{R^2}, \quad \rho = \frac{2}{\kappa \, R^2}, \quad M = \frac{4\pi^2}{\kappa} R,$$

where M is the total mass of the entire space.

In the other possible case, when M is determined by equation (3), we arrive, by an appropriate change of x_4 [10], at de Sitter's spherical world, in which $M = \cos x_1$; using the expression (D2) we find the following relations of de Sitter's world:

$$\lambda = \frac{3c^2}{R^2}, \quad \rho = 0, \quad M = 0,$$

In this way, the stationary world can be *either Einstein's cylindrical world or de Sitter's spherical world.*

2. We will now turn our attention to the study of the other possible world – the non-stationary world. In this case M is a function only of x_4. By suitably changing x_4, we can, without loss of generality, set $M = 1$. Keeping in mind the great convenience of our usual representations, we will write ds^2 in a form that is analogous to (D$_1$) and (D$_2$):

$$ds^2 = -\frac{R^2(x_4)}{c^2}(dx_1^2 + \sin^2 x_1 dx_2^2 + \sin^2 x_1 \sin^2 x_2 dx_3^2) + dx_4^2, \quad (D_3)$$

---

[9] In our case, the density ρ is an unknown function of the world coordinates x_1, x_2, x_3, x_4.

[10] This change is made with the help of the formula: $d\bar{x}_4 = \sqrt{A_0 x_4 + B_0} dx_4$.

Our task is to determine R and ρ from equations (A). It is evident that the equation (A) with differend indices do not give anything, whereas for $i = k = 1, 2, 3$ equations (A) give one relation:

$$\frac{R'^2}{R^2} + \frac{2RR''}{R^2} + \frac{c^2}{R^2} - \lambda = 0. \tag{4}$$

With $i = k = 4$ equation (A) gives the relation:

$$\frac{3R'^2}{R^2} + \frac{3c^2}{R^2} - \lambda = \kappa c^2 \rho, \tag{5}$$

where

$$R' = \frac{\mathrm{d}R}{\mathrm{d}x_4}, \quad R'' = \frac{\mathrm{d}^2 R}{\mathrm{d}x_4^2}.$$

As $R' \neq 0$, the integration of equation (4), after substituting x_4 with t for convenience, gives the equation:

$$\frac{1}{c^2}\left(\frac{\mathrm{d}R}{\mathrm{d}t}\right)^2 = \frac{A - R + \frac{\lambda}{3c^2} R^3}{R} \tag{6}$$

where A is an arbitrary constant. From this equation R can be obtained by inverting an elliptic integral, i.e. by finding R from the equation

$$t = \frac{1}{c}\int_a^R \sqrt{\frac{x}{A - x + \frac{\lambda}{3c^2} x}}\, \mathrm{d}x + B, \tag{7}$$

where B and a are constants; here, of course, the usual change of sign of the square root should be taken care of.

Equation (5) enables us to determine ρ:

$$\rho = \frac{3A}{\kappa R^3}, \tag{8}$$

where the constant A is expressed through the total mass M of space by the following relation:

$$A = \frac{\kappa M}{6\pi^2}. \tag{9}$$

As the mass M is positive, it follows that A is also positive.

3. The study of the non-stationary world is based on the study of equation (6) or (7). Therein, of course, the quantity λ is not determined by itself and in our study of equation (6) or (7) we will assume that λ can take on any value. We will determine those values of the variable x which can change the sign of the square root in equation (7). Restricting ourselves to the case of a positive radius of curvature, it is sufficient to us to consider such values of x for which the quantity under the square root is zero or infinity *in the interval* $(0, \infty)$ *for* x, i.e., for positive x.

One of the values of x, for which the square root in equation (7) becomes zero, is $x = 0$; the other values of x, which can change the sign of the square root in (7), are found by studying the positive roots of the equation

$$A - x + \frac{\lambda}{3c^2} x^3 = 0$$

Denoting $\frac{\lambda}{3c^2}$ by y we can construct a family of third order curves in the (x, y)–plane defined by the equation:

$$y x^2 - x + A = 0, \qquad (10)$$

where A is a parameter of the family which varies in the interval $(0, \infty)$. The curves of the family (see the figure) intersect the x−axis at the point $x = A$, $y = 0$ and have a maximum at the point:

$$x = \frac{3A}{2}, \quad y = \frac{4}{27 A^2}.$$

Examining the figure shows that, for negative λ, the equation

$$A - x + \frac{\lambda}{3c^2} x^3$$

has one positive root x_0, lying in the interval $(0, A)$. Regarding x_0 as a function of λ and A:

$$x_0 = \theta(\lambda, A),$$

we find that θ is an increasing function of λ and an increasing function of A. Further, if λ lies in the interval $(0, 4/9(c^2/A^2))$, then our equation will have two positive roots: $x_0 = \theta(\lambda, A)$ and $x_0' = \vartheta(\lambda, A)$, wherein

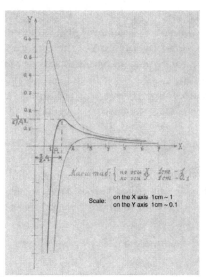

This is the original drawing in Friedmann's manuscript. Image taken from Friedmann's manuscript typed in Russian and preserved in the Ehrenfest archive.

x_0 lies in the interval $(A, 3A/2)$ and x_0' – in the interval $(3A/2, \infty)$; $\theta(\lambda, A)$ will be an increasing function of both λ and A, whereas $\vartheta(\lambda, A)$ will be a decreasing function of λ and of A. At the end, if λ is greater than $\frac{4}{9}\frac{c^2}{A^2}$, our equation will not have positive roots at all.

Beginning our study of formula (7) we will make a remark; at the initial moment, i.e., at $t = t_0$, let the radius of the curvature be R_0. In this initial moment the square root in formula (7) will have a plus or minus sign depending on whether or not the radius of the curvature increases with time at $t = t_0$; by replacing the time t with $-t$ we can always assign a plus sign to this square root; in other words, without loss of generality, we can choose time in such a way that at the initial moment $t = t_0$ the curvature radius increases with time.

4. We consider the case when $\lambda > \frac{4}{9}\frac{c^2}{A^2}$, i.e. the case when the equation

$$A - x + \frac{\lambda}{3c^2} x^3$$

has no positive roots. In this case equation (7) can be written in the following way:

$$t - t_0 = \frac{1}{c} \int_{R_0}^{R} \sqrt{\frac{x}{A - x + \frac{\lambda}{3c^2} x^3}}\, dx, \qquad (11)$$

where according to our remark above the square root will be always positive. It follows from here that *R will be an increasing function of t*; in this case there are no restrictions imposed on the initial value of the curvature radius R_0.

As the curvature radius could not be smaller than zero, by decreasing from R_0 with decreasing of t according to (11), the curvature radius would become zero after a certain period of time t'. Using the obvious analogy, we will call the time interval needed for the radius of curvature to increase from 0 to R_0 — *the time that elapsed since the creation of the world*;[11] this time period t' is determined from the equation:

$$t' = \frac{1}{c} \int_{0}^{R_0} \sqrt{\frac{x}{A - x + \frac{\lambda}{3c^2} x^3}}\, dx, \qquad (12)$$

Let us agree to call, from now on, the world we are discussing *monotonic world of the first kind*.

The elapsed time since the creation of the monotonic world of the first kind, regarded as a function of R_0, A, λ, has the following properties: 1) it increases as R_0 increases; 2) it decreases as A increases, i.e. as the mass in space increases; and 3) it decreases as λ increases. If $A > \frac{2}{3} R_0$, then for any λ the time which elapsed since the creation of the world is finite. If $A \leq \frac{2}{3} R_0$, then there can always be found a characteristic value of $\lambda = \lambda_1 = \frac{4c^2}{9A^2}$ such that as λ approaches this value, the time since the creation of the world would increase indefinitely.

5. We suppose further that λ is confined in the interval $0, \frac{4c^2}{9A^2}$; then the initial value of the curvature radius R_0 can lie in one of the three intervals: $(0, x_0)$, (x_0, x_0'), (x_0', ∞). If R_0 lies in the second interval (x_0, x_0'), then the square root in formula (7) has imaginary value and space with such initial curvature could not exist. The case when R_0 lies in the first interval $(0, x_0)$ will be considered in the next section

---

[11]The time which passed since the creation of the world is in fact the time that elapsed from the moment when space was a point ($R = 0$) to its present state ($R = R_0$); this time might be infinite.

(Section 6). Now we will consider the third case when $R_0 > x'_0$ or $R_0 > \vartheta(\lambda, A)$. In this case, consideration analogous to those in the previous section could show that R would be an increasing function of time, where R could change beginning with $x'_0 = \vartheta(\lambda, A)$. The period of time from the moment when $R = x_0$ to the moment $R = R_0$ will be called the time that elapsed since the creation of the world and will denote it by t':

$$t' = \frac{1}{c} \int_{x'_0}^{R_0} \sqrt{\frac{x}{A - x + \frac{\lambda}{3c^2}x^3}} \, \mathrm{d}x. \tag{13}$$

Let us agree to call the world we are discussing *monotonic world of the second kind*.

6. At the end we consider the case when λ is confined in the interval $(-\infty, 0)$. In this case, if $R_0 > x_0 = \theta(\lambda, A)$ the square root in (7) becomes imaginary and therefore a space with such a curvature radius could not exist. If $R_0 < x_0$, then this case will be identical to the case in the previous section, which we did not discuss there. So let us suppose that λ lies in the interval $(-\infty, \frac{4c^2}{9A^2})$ and that $R_0 < x_0$. We can show in this case, through usual considerations,[12] that R will be a periodic function of t with period $t_\text{п}$, which we call the world period and which will be determined from the expression:

$$t_\text{п} = \frac{2}{c} \int_0^{x_0} \sqrt{\frac{x}{A - x + \frac{\lambda}{3c^2}x^3}} \, \mathrm{d}x. \tag{14}$$

where the curvature radius will change from zero to x_0. We will call such kind of world *periodic*. The period of the periodic world increases as λ increases, approaching infinity when λ approaches $\lambda_1 = \frac{4c^2}{9A^2}$. For small λ the period t_π is given by the approximation formula:

$$t_\text{п} = \frac{\pi A}{c}. \tag{15}$$

---

[12]See, for example, K. Weierstrass, Ueber eine Gattung der reel periodischer Functionen. *Mouatsber. d. Königl. Akad. d. Wissensch.*, 1866, and also J. Horn, Zur Theorie der kleinen endlichen Schwingungen. *Ztschr. f. Mathem. und Physik*, 13d. **47** 1902. The considerations of these authors should be adjusted to our case; in fact the periodicity in our case can be demonstrated through elementary considerations.

The periodic world can be viewed from two perspectives. If we assume that two events coincide, as long as the space coordinates coincide, whereas the time coordinates differ by an integer multiplied by the period, then the radius of curvature of the world increases from 0 to x_0, and after that decrease to zero. In this case the time of existence of the world will be finite.

On the other hand, if time changes from $-\infty$ to $+\infty$, i.e., if we regard two events as coincident as long as not only do their space coordinates coincide, but also their time coordinates coincide, then we arrive at a real periodicity of the space curvature.

7. The experimental evidence at our disposal is completely insufficient for carrying out numerical calculations and for finding out what kind of world is our universe. It might be that the issues of causality and of the centrifugal force could shed light on the questions discussed here. It should be noted that the "cosmological" quantity λ in our formulas is not determined; it is only an extra constant in the formulas. It might be that electrodynamical considerations could determine this quantity. Setting $\lambda = 0$ and assuming that $M = 5 \times 10^{21}$ solar masses, we obtain for the world period a quantity of the order of 10 billion years.

These figures could have, of course, merely illustrative meaning.

Petrograd,
29 May 1922

A. Friedmann,
Professor of Mechanics of the
Petrograd Polytechnic Institute

Image taken from Friedmann's manuscript typed in Russian and preserved in the Ehrenfest archive.

ON THE POSSIBILITY OF A WORLD WITH A CONSTANT NEGATIVE CURVATURE OF SPACE

§1

1. In our Note "On the curvature of space"[1] we considered those solutions of Einstein's cosmological equations, which lead to such models of the world that possess, as a common feature, a constant positive curvature. There we discussed all such possible cases. However, the possibility of obtaining a world of constant positive space curvature from the world equations is closely linked to the finiteness of space. For this reason it is of interest to see whether one can obtain from the same world equations a world of constant negative curvature, about whose finiteness could hardly be talked even with some additional assumptions.

In the present Note it will be shown that it is really possible to obtain a world with constant negative curvature of space from Einstein's cosmological equations. As in the cited work, here too we have to distinguish two cases, namely: 1) the case of a stationary world, whose curvature is constant in time, and 2) the case of a non-stationary world, whose curvature, although constant in space, changes in time. There exist an essential distinction between the stationary worlds of constant negative and of constant positive curvature. Namely the worlds of stationary negative curvature do not allow positive density of matter; the density should be either negative or zero. In relation with this, an

---
[1] A. Friedmann, ZS. f. Phys. 10, 377, 1922, Heft 6.

analog of the physically possible stationary worlds (i.e. worlds with non-negative density of matter) is not Einstein's world, but de Sitter's world.[2]

In the conclusion of this Note we will touch on the question of whether any judgment on the finiteness or infinitness of space can be at all based on the curvature of space.

In the conclusion of this Note we will touch on the question of whether on the grounds of the curvature of space one is allowed at all to judge on its finiteness or infinitude.

2. Let us start with the general assumptions made in the cited Note, which we will divide into the same two classes; we will also keep our notations used there. The assumptions of the first class regard equations (A), (B), (C) of our cited work as Einstein's cosmological equations. The assumptions of the second class will now differ from the ones used in first Note. Assuming that one of the world coordinates, x_4, can be regarded as time coordinate, we can give the second class of assumptions (for the case of the world with negative constant curvature of space) the following formulation: the interval ds^2 should have the form:

$$ds^2 = \frac{R^2(dx_1^2 + dx_2^2 + dx_3^2)}{x_3^2} + M^2 dx_4^2, \qquad (D')$$

where R denotes a function of time and M – a function of all four world coordinates. The constant negative curvature of space of our world is proportional to $-1/R^2$.[3]

Taking into account that for our world ds^2 is an indefinite form, we can, by changing the notation, write formula (D_1) in the following way:

$$d\tau^2 = -\frac{R^2}{c^2} \frac{(dx_1^2 + dx_2^2 + dx_3^2)}{x_3^2} + M^2 dx_4^2. \qquad (D'')$$

Of course, the space curvature of our world remains negative and proportional to $-1/R^2$.

Our task is to find two functions R and M, satisfying Einstein's cosmological equations, i.e. the equations (A), (B) and (C) in the cited Note.

---

[2] That the possibility of a world of a negative curvature of space requires special investigation was pointed out to me by my friend Professor Ya. D. Tamarkin.

[3] Regarding the line element, see, for example, B. Bianchi, Lezioni di geometria differenziale, v. 1. Bologna, 1923, p. 345.

Putting $i = 1, 2, 3$ and $k = 4$ in (A) we obtain the following three equations:

$$R'(x_4)\frac{\partial M}{\partial x_1} = R'(x_4)\frac{\partial M}{\partial x_2} = R'(x_4)\frac{\partial M}{\partial x_3} = 0.$$

These equations show that the worlds we discuss can belong to one of the two types:

Type 1. Stationary worlds, $R' = 0$, R does not depend on time.
Type 1. Non-stationary worlds, $R' \neq 0$, M depends only on time.

Let us consider first the case of the stationary world; the solution for the non-stationary world of negative curvature has a great similarity to the solution for the non-stationary world of constant positive space curvature; that is why we will touch on the non-stationary case quite briefly.

§2

1. For the indices $i, k = 1, 2, 3$ equations (A) gives:

$$\frac{\partial^2 M}{\partial x_1 \partial x_2} = 0, \quad \frac{\partial^2 M}{\partial x_2 \partial x_3} + \frac{1}{x_3}\frac{\partial M}{\partial x_2} = 0, \quad \frac{\partial^2 M}{\partial x_1 \partial x_3} + \frac{1}{x_3}\frac{\partial M}{\partial x_1} = 0.$$

Integrating these equations we obtain:

$$M = \frac{P(x_1, x_4) + Q(x_2, x_4)}{x_3} + L(x_3, x_4), \tag{1}$$

where P, Q and L are for now arbitrary functions of their arguments. The equations (A) serve us for the determination of P, Q and L, where one has to set i,k = 1,2,3. The calculation gives:

$$-\frac{1}{M}\left(\frac{\partial^2 M}{\partial x_2^2} + \frac{\partial^2 M}{\partial x_3^2}\right) = \frac{1 - \lambda R^2}{x_3^2},$$

$$-\frac{1}{M}\left(\frac{\partial^2 M}{\partial x_1^2} + \frac{\partial^2 M}{\partial x_3^2}\right) = \frac{1 - \lambda R^2}{x_3^2}, \tag{2}$$

$$-\frac{1}{M}\left(\frac{\partial^2 M}{\partial x_1^2} + \frac{\partial^2 M}{\partial x_2^2}\right) + \frac{2}{x_3}\frac{1}{M}\frac{\partial M}{\partial x_3} = \frac{1 - \lambda R^2}{x_3^2}.$$

Subtracting the first from the second equation of this system, we obtain:
$$\frac{\partial^2 P}{\partial x_1^2} = \frac{\partial^2 Q}{\partial x_2^2}.$$

It follows from this that:
$$P = n(x_4)x_1^2 + a_1(x_4)x_1 + b_1(x_4),$$
$$Q = n(x_4)x_2^2 + a_2(x_4)x_2 + b_2(x_4). \quad (3)$$

Taking into account (1) and (3) we can write the last equation of (2) in the form

$$-\frac{3-\lambda R^2}{x_3^2}(P+Q) = \frac{4n}{x_3} + \frac{1-\lambda R^2}{x_3^2} - \frac{2}{x_3}\frac{\partial L}{\partial x_3}. \quad (4)$$

So, if $P+Q$) really contains one of the quantities x_1 or x_2, i.e., if one of the coefficients n, a_1, a_2 is different from zero, then as the right-hand side of this equation does depend either on x_1 or x_2, the multiplier of $P+Q$ in equation (4) must be zero. The case, when all of the three quantities n, a_1, a_2 become zero, should be considered separately.

In such a way, in the case when not all quantities n, a_1, a_2 are zero, λ and the curvature of space are linked by the relation:

$$\lambda R^2 = 3. \quad (5)$$

Taking into account (5), equations (2) reduce to a single equation which defines the function L, namely:

$$\frac{\partial L}{\partial x_3} + \frac{L}{x_3} = 2n \quad (6)$$

2. Below we should distinguish two cases: 1) $n \neq 0$, and 2) $n = 0$. In the first case, as shown by equations (D$_1$), (1), (3), it is possible, without loss of generality, to assume that the quantity n is equal to 1; namely, using the substitution $x_4 = \varphi(x_4)$ we can always have $n = 1$. Using this, we obtain from (6):

$$L = \frac{L_0(x_4)}{x_3} + x_3. \quad (7)$$

In order to determine ρ, we put $i = k = 4$ in equations (A); a simple calculation shows that in our case ρ becomes zero. Therefore, the first case is characterised by zero density of matter and by the interval:

$$ds^2 = \frac{R^2}{x_3^2}(dx_1^2 + dx_2^2 + dx_3^2)$$
$$+ \left[\frac{x_1^2 + x_2^2 + a_1(x_4)x_1 + a_2(x_4)x_2 + a_3(x_4) + x_3^2}{x_3}\right]^2 dx_4^2. \quad (D_1')$$

Turning to the second case ($n = 0$), we find for L the equation:

$$L = \frac{L_0(x_4)}{x_3}. \quad (8)$$

In this case also, the calculation gives a zero value for ρ. Therefore, the second case characterizes by zero density of matter and by the interval as well:

$$ds^2 = \frac{R^2}{x_3^2}(dx_1^2 + dx_2^2 + dx_3^2)$$
$$+ \left[\frac{a_1(x_4)x_1 + a_2(x_4)x_2 + a_3(x_4)}{x_3}\right]^2 dx_4^2. \quad (D_2')$$

At the end, let us consider the case when all three coefficients n, a_1, a_2 are zero, which means that M does not depend on x_1 and x_2. By integrating (2) we again arrive at two cases:

1) $\lambda R^2 = 3$, $M = \dfrac{M_0(x_4)}{x_3}$,

2) $\lambda R^2 = 1$, $M = M(x_4)$,

where M_0 and M are arbitrary functions of their arguments. The interval for the first case is a special case of formula (D_2'); the previous calculation shows that the density of matter here is zero.

The second case[4] leads, as is easily seen, to a density of matter, which is different from zero. Then, to decide whether the density would

---
[4] W. Fock pointed out to me that such case was possible.

be positive or negative, it is necessary to use that form of the interval, which corresponds to an indefinite quadratic form, and is expressed by (D''). Carrying out calculations with the gravitational potentials of (D''), we see that in the considered case M is a function only of x_4. Therefore, without loss of generality, it is possible to put $M = 1$ (for this, it is only necessary to use the coordinate $x_4 = \varphi(x_4)$ instead of x_4). With this assumption we calculate the density ρ and find

$$\lambda = -\frac{c^2}{R^2}, \qquad \rho = \frac{2}{\kappa R^2}, \qquad \text{(D}_3''\text{)}$$
$$d\tau^2 = -\frac{R^2}{c^2} \frac{dx_1^2 + dx_2^2 + dx_3^2}{x_3^2} + dx_4^2.$$

Therefore, this case gives a negative value of ρ.

Summarising, it can be said that a *stationary world* with constant negative curvature of space is possible only for zero or negative density of matter; the interval corresponding to this world is given by the above formulas (D$_1'$), (D$_2'$) and (D$_3''$).

3. Let us now consider the case of a non-stationary world. First of all, let us note that here M is a function only of x_4; considerations, which we discussed not once earlier, show that M can be equated to unity. With these assumptions, we can easily find that equations (A) for $i = 1, 2, 3$, $k = 4$ and for $i \neq k$, $i, k = 1, 2, 3$ are identically satisfied. Setting $i = k = 1, 2, 3$ in (A), we obtain a differential equation of second order, which determines the function $R(x_4)$, namely:

$$\frac{R'^2}{R^2} + \frac{2RR''}{R^2} + \frac{1}{R^2} - \lambda = 0. \qquad (9)$$

This equation is completely analogous to our previous equation (equation (4) of the cited work); that equation coincides exactly with (9), if $c = 1$ there. Therefore, the whole discussion of equation (4) can be switched to the above equation (9). That is why we will not go into detail, and will calculate the density of matter ρ for the non-stationary world.

Writing the interval for the non-stationary in the form of (D''), we obtain the differential equation for R:

$$\frac{R'^2}{R^2} + \frac{2RR''}{R^2} - \frac{c^2}{R^2} - \lambda = 0.$$

The integration of this equation gives us the relation:
$$\frac{R'^2}{c^2} = \frac{A + R + \frac{\lambda}{3c^2}R^3}{R},$$
where A is an arbitrary constant. Calculating the density of matter, we get
$$\rho = \frac{3A}{\kappa R^3}. \tag{10}$$
Formula (10) shows that for a positive A the density of matter is also positive.

It follows from this that *non-stationary worlds with constant negative curvature of space and with positive density of matter could exist.*

§3

1. Let us now discuss the physical meaning of the result obtained in the preceding paragraphs. We saw that the Einstein's cosmological equations have solutions describing a world with constant negative curvature of space. This fact shows that the cosmological equations alone are not sufficient to answer the question of the finiteness of our world. Knowledge of the curvature of space does not yet give us direct indications about its finiteness or infiniteness. In order to arrive at a definite conclusion on the finiteness of space, it is necessary to make some additional clarifications. In fact we call a space finite, if the distance between two arbitrary non-coincident points does not exceed a certain positive constant number, no matter what a pair of points that might be. Therefore, before we consider the issue of the finiteness of space, we should agree on which points of this space we should regard as different. For example, if we going to regard the sphere as a surface of the three-dimensional Euclidean space, then the points which lie on the same parallel with a difference in longitude of $360°$, we consider coincident. On the contrary, if we considered these points as different, we would obtain a multi-sheeted spherical surface in Euclidean space. The distance between two arbitrary points on a sphere does not exceed a given finite number. If we regard this sphere as an infinitely multi-sheeted surface, then this distance can be made arbitrarily large (appropriately comparing the points of different sheets). It is clear from here that before discussing the finiteness of the world, it should

be clarified what points should be regarded as coincident and what as different.

2. As a criterion for a non-coincidence of points could serve, together with others, the principle "fear of ghosts." By this we mean the axiom: between two different points one and only one straight (geodesic) line can be drawn. Accepting this principle, two points which can be joined by more than one straight line can no longer be regarded as different. For example, due to this principle the two end points of the same diameter of a sphere will not differ from each other. Of course, this principle excludes the possibility of a ghost, since a ghost appears at the same point as the image itself which is creating it.

The just considered definition of the notion of coincident and non-coincident points leads to the understanding that spaces with positive constant curvature are finite. However, the criterion mentioned above does not allow us to make any conclusion about the finiteness of spaces of negative constant curvature. This gives grounds to assert Einstein's cosmological equations alone, without additional assumptions are not yet sufficient to arrive at a conclusion about the finiteness of our world.

Petrograd, November 1923.

ON THE GEOMETRY OF CURVED SPACES

In his known book "Raum, Zeit, Materie" Weyl outlined the foundations of the geometry of curved spaces by employing a notion of a parallel transport of a vector that is more general than the one developed by Levi-Civita,[1] and attaches to this notion a new and extremely original idea of a metric connection of space. The development of these geometrical ideas enabled Weyl to generalize Einstein's ideas and to obtain the derivation of Maxwell's equations (together with Mie's theory) from the geometrical properties of space.

By generalizing the idea of a parallel transport, Weyl introduced a number of essential restrictions. For example, he assumed a symmetry in lower indices of the quatities $\Gamma^i_{\lambda\mu}$ and also linked in a certain way the parallel transport of covariant and contravariant vectors. On the other hand, the geometrical reason of why the change of the norm of a vector under parallel transport is proportional to the norm of the vector, in Weyl's presentation, is not clear.

Taking into account the remarkable results of Weyl, it seems interesting to get rid of the restrictions mentioned above, and also to clarify the geometrical reason for the special kind of metric connection of space which is employed by Weyl.[2]

The generalization of the geometrical ideas of Weyl should first of all get rid of the link imposed on the parallel transport of a con-

---

[1] See Levi-Civita, Nozione di parallelismo in una varieta qualunque etc., Rendic. del Circolo Matem. di Palermo, t. 42 (1917).

[2] A generalization in a certain direction of Weyl's space was made by Eddington in his article published in Proceed. Lond. Roy. Soc. Vol. 1921.

travariant vector making it dependable on the parallel transport of the covariant vector. It looks useful while getting rid of this restriction to classify the spaces obtained in this way, by considering the geometrical objects attached to these spaces. Such a geometrical object will be the notion of a plane, correspondingly generalized for the case of a curved space of n dimensions. Such a classification makes it possible to select from the spaces considered, such spaces where Weyl's space will be a special case.

The geometrical reason for the form of the metric connection which is employed by Weyl, as will be shown later, consists in that this form is a necessary and sufficient condition for the angles of covariant vectors not to change under parallel transport of vectors. Imposing such a requirement for contravariant vectors, we can find the parameters determining the parallel transport of both co- and contravariant vectors with the help of the fundamental metric tensor g_{ik}, with two (not one as done by Weyl) contravariant scale vectors, and with two special tensors of third rank. Concentrating on spaces, in which these tensors of third rank become zero or depend on both the metric tensor and the two scale vectors, we will be able to determine the properties of space with help of the fundamental metric tensor and the two contravariant scale vectors.

For such a kind of generalized space the number of coordinate and scale invariants (in Weyl's meaning) will be much greater than for Weyl's space. Some of them, which are analogous to Weyl's invariants, are not difficult to construct. Thus a conclusion looks self-evident – could it be possibly to obtain Maxwell equations from the properties of the more general space, without accepting Mie's theory? At the end of the present note I will outline some ideas on this issue. However, I think it is necessary to point out here that the presence of one more contravariant scale vector must with absolute necessity lead to a system of equations supplementary to Maxwell's equations. This system of equations should, as it seems, be some kind of generalization of Mie's theory.

§1

1. Let the n-dimensional manifold M_n have as variables (coordinates) its $x_1, x_2, \ldots, x_n$. Replacing these coordinates, through a point

transformation, with new coordinates $\overline{x}_1, \overline{x}_2, \ldots, \overline{x}_n$, we form the manifold $\overline{M}_n$, which we will refer to as *obtained from M_n through a point transformation*. From now on, overlined quantities will indicate that those quantities result from transformations of M_n to $\overline{M}_n$.

If for each of the manifolds M_n, transforming themselves into one another through point transformations of the coordinates, we will have a system of n^3 functions of these coordinates $\Gamma^i_{\lambda\mu}$, transforming in accordance with:

$$\overline{\Gamma}^i_{\lambda\mu} = \Gamma^\gamma_{\alpha\beta} \frac{\partial x_\alpha}{\partial \overline{x}_\lambda} \frac{\partial x_\beta}{\partial \overline{x}_\mu} \frac{\partial \overline{x}_i}{\partial x_\gamma} + \frac{\partial^2 x_\nu}{\partial \overline{x}_\lambda \partial \overline{x}_\mu} \frac{\partial \overline{x}_i}{\partial x_\nu}, \tag{1}$$

then we will call the quantities $\Gamma^i_{\lambda\mu}$ *tensorial parameters*.

From equation (1) it is easy to obtain the following relation:

$$\frac{\partial^2 x_\nu}{\partial \overline{x}_\lambda \partial \overline{x}_\mu} = \overline{\Gamma}^\beta_{\lambda\mu} \frac{\partial x_\nu}{\partial \overline{x}_\beta} - \Gamma^\nu_{\alpha\beta} \frac{\partial x_\alpha}{\partial \overline{x}_\lambda} \frac{\partial x_\beta}{\partial \overline{x}_\mu} \tag{2}$$

Using Christoffel brackets, formed with the help of the symmetric tensor g_{ik}, makes it clear that any tensorial parameters are defined according to the formula:

$$\Gamma^i_{\lambda\mu} = \left\{ \begin{matrix} \lambda\mu \\ i \end{matrix} \right\} + A^i_{\lambda\mu} \tag{3}$$

where $A^i_{\lambda\mu}$ is an arbitrary mixed tensor of third rank.

2. In the study of the properties of tensorial parameters a significant role is played by two tensors of a third and forth rank, which are defined by the equations:

$$\gamma^i_{\lambda\mu} = \Gamma^i_{\lambda\mu} - \Gamma^i_{\mu\lambda}, \tag{4}$$

$$F^i_{k\lambda\mu} = \frac{\partial \Gamma^i_{k\lambda}}{\partial x_\mu} - \frac{\partial \Gamma^i_{k\mu}}{\partial x_\lambda} + \Gamma^\sigma_{k\lambda} \Gamma^i_{\sigma\mu} - \Gamma^\sigma_{k\mu} \Gamma^i_{\sigma\lambda}. \tag{5}$$

That these expressions are tensors follows from equations (1) and (2) constituting transformations of tensorial parameters.

The first of the above tensors, which we introduced, $\gamma^i_{\lambda\mu}$ turns identically zero for the cases considered by Weyl. We will call this tensor symmetral. Tensor $F^i_{k\lambda\mu}$ is called the *curvature of tensorial parameters*. If the symmetral is zero, we will call the tensorial parameters *symmetric*.

It is not difficult to verify that the curvature of tensorial parameters satisfies the relation:
$$F^i_{k\lambda\mu} = -F^i_{k\mu\lambda}. \tag{6}$$

By combining the curvature with the fundamental tensor g_{ik} and with the unity tensor δ^k_i we will have the following tensors and scalars:
$$\begin{aligned} F_{iklm} &= g_{\sigma i}\, F^\sigma_{klm}, \\ F_{ik} &= F^\sigma_{i\sigma k} = F^\sigma_{isk}\,\delta^s_\sigma = g^{\alpha\beta}\,F_{\alpha i\beta k}, \\ F &= g^{ik}\,F_{ik} = g^{ik}\,g^{\alpha\beta}\,F_{\alpha i\beta k}, \end{aligned} \tag{7}$$

We will call the first of these tensors the Riemann tensor, the second – the contracted Riemann tensor, and the third – the scalar curvature.

In the case when $\Gamma^i_{\lambda\mu} = \left\{\begin{matrix}\lambda\mu \\ i\end{matrix}\right\}$, the tensors and scalars defined by (5) and (7), turn into the usual symbols of Riemann or tensors formed from them by contraction.

3. The parallel transport of a covariant vector ξ^i is defined by Weyl in the following way:

Let us consider the curve $\mathcal{K} : x_s = x_s(t)$ and also a covariant vector ${}^o\xi^i$ at a point $(t = t_0)$ of the curve. We will regard that this vector is parallel transported along the curve $\mathcal{K}$, if at any one of its points it is defined as a solution of the system of linear differential equations:
$$\frac{d\xi^i}{dt} = -\Gamma^i_{rs}\,\xi^r\,\frac{dx_s}{dt} \tag{8}$$
with the following initial conditions:
$$t = t_0, \quad \xi^i = {}^o\xi^i,$$
where Γ^i_{rs} will be tensorial parameters.

It is not difficult to see that for the manifold $\overline{M}_n$ the equations (8) can be written as:
$$\frac{d\overline{\xi}^i}{dt} = -\overline{\Gamma}^i_{rs}\,\overline{\xi}^r\,\frac{d\overline{x}_s}{dt}$$

i.e. they preserve their form.

The parallel transport of a covariant vector along a closed curve (for arbitrary tensorial parameters Γ^i_{rs}), generally speaking, does not bring the vector to its initial position when it is returned in the initial point. In other words, the parallel transport of a vector *will depend on the path, along which the transport is carried out.*

Weyl proved a theorem that *a necessary and sufficient condition for the independence of the parallel transport of a vector of the path, along which this transport takes place, is the zero value of the curvature tensorial parameters* $\Gamma^i_{\lambda\mu}$.

For symmetric tensorial parameters the just mentioned condition makes it possible by a coordinate transformation to bring all the tensorial parameters to zero for any points of M_n. For non-symmetric parameters this cannot be stated.

We will define parallel transport of contravariant vectors along the curve $\mathcal{K}$ in the following way. Let the contravariant vector $^o\eta_i$ be defined at some point ($t = t_0$) of this curve. We will say that this vector is parallel transported along the curve $\mathcal{K}$, if at any point of the curve it is defined as a solution of the system of differential equations:

$$\frac{d\eta_i}{dt} = G^r_{is}\,\eta_r\,\frac{dx_s}{dt}, \tag{9}$$

under the following initial conditions:

$$t = t_0, \quad \eta^i = {}^o\eta_i,$$

where $G^i_{\lambda\mu}$ are tensorial parameters. We will call the tensorial parameters $\Gamma^i_{\lambda\mu}$ *covariant*, and the parameters $G^i_{\lambda\mu}$ – *contravariant tensorial parameters*. In Weyl's geometries these two kinds of parameters coincide. It is not difficult to see that the expression $\alpha = \xi^i \eta_i$ does not change as long as ξ^i and η_i will be parallel transported along the same curve. In fact simple calculations give:

$$\frac{d\alpha}{dt} = \xi^i\,\eta_r\,\frac{dx_s}{dt}\,(G^r_{is} - \Gamma^r_{is}),$$

from where our conclusion follows.

However, we do not assume that the two kinds of tensorial parameters coincide. Below, on the basis of geometrical considerations, we will establish some connection between them which is more general assumption than the one of Weyl.

The same conclusions about parallel transport of contravariant vectors can be made about covariant vectors as well. A necessary and sufficient condition for the independence of the parallel transport of a contravariant vector on the path, along which the transport is carried out, is the zero value of the curvature contravariant tensorial parameters.

4. It is completely clear that regarding space from the viewpoint of covariant or contravariant vectors forming this space, and taking into account the notion of parallel transport defined above, and describing space with some manifold M_n, we define its vector properties with covariant and contravariant vectorial parameters. This general vector space allows for a simple classification, in terms of both of covariant and contravariant vectors. We will consider the clasification only of covariant vectors. The general space divides into two classes: *non-symmetric*, when the symmetral is non-zero and *symmetric*, when the symmetral is zero. The symmetric space itself divides into two classes:

1) *general symmetric*, when $\Gamma^i_{\lambda\mu} = \left\{ {\lambda\mu \atop i} \right\} + A^i_{\lambda\mu}$, where $A^i_{\lambda\mu}$ is a tensor, which is symmetric in the lower indices, and is different from zero;

2) *Riemannian*, when $\Gamma^i_{\lambda\mu} = \left\{ {\lambda\mu \atop i} \right\}$. Among the general symmetric spaces we will later select a class of spaces, where the space considered by Weyl; we call this class of spaces *Weyl's spaces*. Among the Riemannian spaces we will select a special class of spaces, where all $\Gamma^i_{\lambda\mu}$ are zero; we will call these spaces *Euclidean*, having in mind that with the help of point transformations always can choose such a manifold M_n describing the space in which all g_{ik} will be constant quantities. As a matter of fact, these will be, of course, rather pseudo-Euclidean, not Euclidean, due to the law of inertia of quadratic forms.

§2

1. We now turn to the definition of the notions of a straight line, a plane, and to the clarifications of some basic properties of these geometric objects.

The *direction* of a covariant or contravariant vector is defined by $n-1$ relations of its components.

Obviously the components of two vectors, having the same direction, differ only by a factor, which is the same for all components. For example, if a^i and b^i are two vectors having the same direction, then $a^i = \lambda b^i$ for all i from 1 to n.

We will say that *the direction of the vector ξ^i (or η^i) is parallel transported along the curve $\mathcal{K}$*, if it is possible to find such a function λ of the points of the curve, so that the vector $\lambda \xi^i$ (or $\lambda \eta_i$) could be parallel transported along the curve $\mathcal{K}$.

It is not difficult to show that *a necessary and sufficient condition for the direction of a covariant vector ξ^i to be parallel transported along the curve $\mathcal{K}$ is the independence of the index i of the relation:*

$$\frac{\frac{d\xi^i}{dt} + \Gamma^i_{rs} \xi^r \frac{dx_s}{dt}}{\xi^i} = \lambda.$$

In the same way, *the necessary and sufficient condition for the direction of a contravariant vector to be parallel transported along the curve $\mathcal{K}$ is the independence of the index i of the following relation:*

$$\frac{\frac{d\eta_i}{dt} - G^r_{is} \eta_r \frac{dx_s}{dt}}{\eta_i} = \mu.$$

2. We will call *the direction of the tangent to the curve $x_i = x_i(t)$* at a given point, the direction of the covariant vector $\frac{dx_i}{dt}$.

It is not difficult to see that the direction of the tangent to the curve is completely independent of the parameter with which the curve is parametrized.

We will call *a straight line the curve, the direction of whose tangent is parallel transported along the curve.*

It follows from the above that the equation of a straight line reduces to the following system of $n-1$ differential equations of the second order:

$$\frac{\frac{d^2 x_i}{dt^2} + \Gamma^i_{rs} \frac{dx_r}{dt} \frac{dx_s}{dt}}{\frac{dx_i}{dt}} = \lambda,$$

where λ is an undefined function of t.

It is not difficult to see that the parameter t can be always chosen in such a way that the equation of the straight line would be written in the form:
$$\frac{d^2 x_i}{dt^2} + \Gamma^i_{rs} \frac{dx_r}{dt} \frac{dx_s}{dt} = 0 \qquad (10)$$
We will call this kind of parameter t *principal*, since in the definition of a straight line it plays the same role as the length of the arc in the definition of the equation of a geodesic.

A closer look at equation (10) shows that only one line can be drawn through a given point at which the direction of the tangent is also given. In the case when our space does not belong to the Riemannian spaces, then the straight will not be geodesic and only for Riemannian spaces the notions of geodesic and straight lines coincide.

3. We arrived at the concept of a straight line by considering covariant vectors and their parallel transport; a definition of a flat hypersurfaces can be obtained by employing the notion of parallel transport of a contravariant vector.

We will call the contravariant normal to the hypersurface S : $f(x_1, x_2, \ldots, x_n) = 0$ at a given point, a contravariant vector:
$$f_i = \frac{\partial f}{\partial x_i}$$
at this point.

For any curve lying on S will have:
$$f_i \frac{dx_i}{dt} = 0.$$
and conversely, if the equation
$$a_i \frac{dx_i}{dt} = 0.$$
will hold for any curve lying on S, then the direction of the vector a_i and the direction of the contravariant normal to S at the given point coincide.

A plane in the ordinary Euclidean space has the property that the normal to the plane is parallel transported along any curve lying on the plane.

Searching for hypersurfaces in a general space possessing the property, that the direction of the contravariant normal is parallel transported along any curve lying on the hypersurface, we arrive at very restrictive constraints for the curvature of the covariant tensorial parameters, constraints making the space very similar to the Euclidean, where the curvature of the covariant tensorial parameters becomes zero. For lack of space we will not discuss this question any further.

A plane in the usual Euclidean space possesses also the following property, which is more convenient for the transition to the general space.

If from a given point of a plane in the usual Euclidean space we draw a bunch of straight lines, lying on the plane, then the direction of the normal to the plane is parallel transported along each of these straight lines.

To each point of space corresponds a direction of a normal and a plane which is perpendicular to this direction, in the usual Euclidean space, a plane drawn, through a given point, perpendicular to a given direction, will be, of course, a plane for any other of its points drawn perpendicularly to the normal at this point; the generalized notion of plane hypersurface does not have this property.

We will call the straight hypersurface at a given point P, which is perpendicular to the contravariant vector $^{o}f_i$, a hypersurface formed by all straight lines passing through this point and having at point P a given contravariant normal whose directions coincides with the direction of the vector $^{o}f_i$.

It is not difficult to see that the straight hyperspace at point P is perpendicular to the vector $^{o}f_i$ always exists and is defined in a unique way.

We will call the straight hypersurface at point P, perpendicular to the vector $^{o}f_i$, a flat hypersurface corresponding to the point P and perpendicular to the vector $^{o}f_i$, if the contravariant normal to this straight hypersurface is parallel transported along any straight line going through point P.

It is completely clear that a flat hypersurface, corresponding to point P, will not be not a flat hypersurface corresponding to any other point lying on it; the flat hypersurface has this kind of property only in extraordinary cases.

We will call a space, in which each straight hypersurface is a flat hypersurface, a *vectorially perfect space*.

3. Let us clarify the requirements imposed on covariant and contravariant tensorial parameters in order that a space be vectorially perfect.

Theorem. *A necessary and sufficient condition for a space the to be vectorially perfect is: 1) the contravariant tensorial parameters are symmetric and 2) that the following relations are satisfied:*

$$\iota^i_{rs} + \iota^i_{sr} = 0, \ (i \text{ not} = r, \ i \text{ not} = s),$$
$$\iota^i_{ri} = \iota^i_{ir} = \omega_r, \text{ does not depend on } i, \ (i \text{ not} = r), \qquad (11)$$
$$\iota^i_{ii} = \omega_i{}^3$$

where $\iota^i_{\lambda\mu} = G^i_{\lambda\mu} - \Gamma^i_{\lambda\mu}$.

It is not difficult to see that $\iota^i_{\lambda\mu}$ is a mixed tensor of third rank and ω_i is a contravariant vector. Equations (11) can be written in following way:

$$\iota^i_{rs} + \iota^i_{sr} = \delta^i_r \omega_s + \delta^i_s \omega_r. \qquad (12)$$

We will precede the proof of this theorem with two lemmas.

Lemma 1. *If a_{rs} does not depend on ξ^i, and if $a_{rs} \xi^r \xi^s = 0$ for all ξ^i satisfying the condition $f_i \xi^i = 0$, then:*

$$a_{rs} + a_{sr} = f_s \frac{a_{rr}}{f_r} + f_r \frac{a_{ss}}{f_s} \qquad (a)$$

It follows from the conditions of the lemma that the quadratic form $a_{rs} \xi^r \xi^s$ can be divided by the linear form $f_i \xi^i$ i.e. the following identity holds:

$$a_{rs} \xi^r \xi^s = (f_i \xi^i)(\lambda_j \xi^j).$$

By comparing the coefficients before $\xi^r \xi^s$ on both sides of this identity, we prove the lemma.

Lemma 2. *If a^i_{rs} does not depend on f_i and ξ^i, and if $a^i_{rs} f_i \xi^r \xi^s = 0$ for all f_i and ξ^i satisfying the condition $f_i \xi^i = 0$, then a^i_{rs} satisfies the relation:*

$$a^i_{rs} + a^i_{sr} = \delta^i_r \omega_s + \delta^i_s \omega_r, \qquad (b)$$

where $\omega_r = a^r_{rr}$.

---

[3]Of course, there is no summation over i.

Using formula (a) of lemma 1 we will have for all f_i the following equality:
$$a^i_{rs} f_i + a^i_{sr} f_i = f_s \frac{a^i_{rr} f_i}{f_r} + f_r \frac{a^i_{ss} f_i}{f_s}.$$

By eliminating the denominator we obtain:[4]
$$(a^i_{rs} + a^i_{sr}) f_i f_s f_r = f_s^2 f_i a^i_{rr} + f_r^2 f_i a^i_{ss}.$$

Comparing the coefficient before the different products $f_i f_s f_r$ on both sides of the equality, we prove the lemma.

Now let us prove the necessary condition of our theorem. Let $f(x_1, x_2, \ldots, x_n) = 0$ be a flat hypersurface at point P, and let $f_i = \frac{\partial f}{\partial x_i}$, and $\xi^s = \frac{dx_s}{dt}$, if $x_s = x_s(t)$ is the equation of one of the straight lines going through P and lying on the flat hypersurface. For each of these straight lines we will have:
$$f_i \xi^i = 0.$$

Regarding t as the principal parameter of some straight line, differentiating the preceding equality by t and keeping in mind that for the straight line:
$$\frac{d\xi^i}{dt} = -\Gamma^i_{rs} \xi^r \xi^s,$$

we obtain the following relation:
$$\xi^r \left(\frac{df_r}{dt} - \Gamma^i_{rs} f_i \xi^s \right) = 0. \qquad (c)$$

At the same time keeping in mind, that by the property of a flat hypersurface, the direction of the normal to it is parallel transported along the straight line forming this plane, we can find the following relation:
$$\frac{df_r}{dt} - G^i_{rs} f_i \xi^s = \omega' f_r. \qquad (d)$$

Multiplying this equation by ξ^r and summing over r from 1 to n, and subtracting the obtained result from equation (c), we find:
$$\iota^i_{rs} f_i \xi^r \xi^s = 0$$

---
[4] The summation is only over i.

which relation holds for a vectorial perfect space for all f_i, ξ^i satisfying the condition $f_i \xi^i = 0$. Direct employment of lemma 2 gives us the second condition of the theorem.

In order to obtain the first conditions, note that

$$\frac{df_r}{dt} = \frac{\partial f_r}{\partial x_s} \xi^s,$$

then we can easily obtain the following relation from (d):

$$\left(\frac{\partial f_r}{\partial x_s} - G^i_{rs} f_i\right) \xi^s = \omega' f_r.$$

Multiplying this relation by ξ^r, summing over r from 1 to n and denoting by a_{rs} the expression

$$\frac{\partial f_r}{\partial x_s} - G^i_{rs} f_i$$

we find:

$$a_{rs} \xi^r \xi^s = 0$$

for $f_i \xi^i = 0$. Using lemma 1 we find from here:

$$a_{rs} + a_{sr} = f_s \frac{a_{rr}}{f_r} + f_r \frac{a_{ss}}{f_s}.$$

Multiplying this equation by ξ^s and summing over s from 1 to n we find:

$$a_{rs} \xi^s + a_{sr} \xi^s = f_r \sum_{s=1}^{n} \frac{a_{ss}}{f_s} \xi^s = \omega'' f_r,$$

where ω'' does not *depend* on r. By the definition of a_{sr} we have:

$$a_{sr} = a_{rs} + (G^i_{rs} - G^i_{sr}) f_i.$$

Therefore the preceding equation can be written in this way:

$$2 a_{rs} \xi^s + (G^i_{rs} - G^i_{sr}) f_i \xi^s = \omega'' f_r,$$

but

$$a_{rs} \xi^s = \omega' f_r$$

because:
$$(G^i_{rs} - G^i_{sr})\xi^s f_i = \omega f_r, \quad \omega = \omega'' - 2\omega'.$$

From this equation we find:
$$\frac{(G^i_{rs} - G^i_{sr})\xi^s f_i}{f_r} = \frac{(G^i_{\alpha s} - G^i_{s\alpha})\xi^s f_i}{f_\alpha}$$

Eliminating the denominator and comparing the coefficients before some of the products $f_i f_\beta$ we find:
$$(G^i_{rs} - G^i_{sr})\xi^s = 0, \quad i \text{ not} = r.$$

Since these equations hold for all ξ^i satisfying the relation $f_i \xi^i = 0$ and since $G^i_{rs} - G^i_{sr}$ does not depend on f_i, then:
$$G^i_{rs} - G^i_{sr} = 0, \quad i \text{ not} = r.$$

Setting $i = s$ we find:
$$G^s_{rs} - G^s_{sr} = 0$$

These equation proves the symmetry of the contravariant tensorial parameters.

Let us now prove the sufficient condition of the theorem.

It follows from the condition $f_i \xi^i = 0$ for $\xi^i = \frac{dx_i}{dt}$, where t is the principal parameter of the straight line lying on the flat hypersurface $f = 0$ and passing through point P, that:
$$\frac{df_i \xi^i}{dt} = 0$$

or
$$\xi^r \left(\frac{df_r}{dt} - \Gamma^i_{rs} f_i \xi^s \right) = 0.$$

Taking into account that
$$\Gamma^i_{rs} = G^i_{rs} - \iota^i_{rs}$$

we find:
$$\xi^r \left(\frac{df_r}{dt} - G^i_{rs} f_i \xi^s \right) + \iota^i_{rs} f_i \xi^r \xi^s = 0,$$

but due to the conditions of the theorem we have:
$$\iota^i_{rs} f_i \xi^r \xi^s = \tfrac{1}{2}(\iota^i_{rs} + \iota^i_{sr}) f_i \xi^r \xi^s = \tfrac{1}{2}(\delta^i_r \omega_s + \delta^i_s \omega_r) f_i \xi^r \xi^s = 0,$$
because $f_r \xi^r = 0$; i.e.:
$$\xi^r \left(\frac{\mathrm{d} f_r}{\mathrm{d} t} - G^i_{rs} f_i \xi^s \right) = a_{rs} \xi^r \xi^s = 0, \tag{e}$$
where
$$a_{rs} = \frac{\mathrm{d} f_r}{\mathrm{d} x_s} - G^i_{rs} f_i$$
is symmetrical in the lower indices due to the conditions of the theorem:
$$a_{rs} = a_{sr}.$$
Keeping in mind that equation (e) holds for all ξ^i, satisfying the condition $f_i \xi^i = 0$, and using lemma 1 we find:
$$a_{rs} + a_{sr} = 2 a_{rs} = f_s \frac{a_{rr}}{f_r} + f_r \frac{a_{ss}}{f_s}.$$
Multiplying this equation by ξ^s and summing over s from 1 to n we get the following relation:
$$a_{rs} \xi^s = f_r \cdot \tfrac{1}{2} \sum_{s=1}^{n} \frac{a_{ss}}{f_s} \xi^s.$$
In other words:
$$a_{rs} \xi^s = \frac{\mathrm{d} f_r}{\mathrm{d} t} - G^i_{rs} f_i \xi^s = \mu f_r,$$
i.e. the contravariant normal to the considered straight hypersurface is indeed parallel transported along the straight lines going through point P and lying on the hypersurface. Therefore, the considered straight hypersurface is a flat hypersurface and, in this way, the sufficient condition of the theorem is proven.

It follows directly from equation (12) and from the symmetry of the contravariant parameters that they can be expressed through the covariant parameters:
$$G^i_{\lambda\mu} = \frac{\Gamma^i_{\lambda\mu} + \Gamma^i_{\mu\lambda}}{2} + \tfrac{1}{2}\delta^i_\lambda \omega_\mu + \tfrac{1}{2}\delta^i_\mu \omega_\lambda \tag{13}$$

where ω_λ is any contravariant vector.

In the case of the symmetry of covariant tensorial parameters equation (13) gives the following expression:

$$G^i_{\lambda\mu} = \Gamma^i_{\lambda\mu} + \tfrac{1}{2}\delta^i_\lambda \omega_\mu + \tfrac{1}{2}\delta^i_\mu \omega_\lambda. \tag{14}$$

The spaces considered by Weyl ($G^i_{\lambda\mu} = \Gamma^i_{\lambda\mu}$) can be obtained for the case when $\omega_\lambda = 0$. We will call such kind of spaces where $\omega_\lambda = 0$ and $\Gamma^i_{\lambda\mu}$ are symmetric, *principal vectorially perfect spaces*. We will call the vector ω_i *characteristic vector*.

The consideration of flat hypersurfaces leads us, in this way, to a completely natural definition of a contravariant vector ω_i, which is different from Weyl's scale vector. Thus we have an indication that the properties of space in addition to the metric tensor g_{ik} and the scale vector φ_i is determined by one more vector ω_i. If the properties of space are expressed in the equations of gravitation and electrodynamics, then it is possible that this new vector might be interpreted as one of the vectors entering the equations of electrodynamics.

§3

1. To consider the metric properties of space we introduce the following notations:
The norm of a covariant vector ξ^i, according to Weyl's definition, will be called the following scalar:

$$l = l(\xi^i) = g_{ik}\,\xi^i\,\xi^k, \tag{15}$$

where g_{ik} is the fundamental tensor.
The norm of a contravariant vector η_i will be called the scalar formed in the following way:

$$L = L(\eta_i) = g^{ik}\,\eta_i\,\eta_k, \tag{16}$$

where g^{ik} is a conjugate tensor of the fundamental tensor g_{ik}.

Our immediate task will be to study the change of the norms of the covariant and contravariant vectors when these vectors are parallel transported along some curve $\mathcal{K}$.

As we indicated above, Weyl considers only such types of metrical spaces, in which the norm of the covariant vector, when it is parallel transported along the curve $x_i = x_i(t)$, changes according to the following law:

$$\frac{dl}{dt} = -\varphi_s l \frac{dx_s}{dt}$$

where φ_s is a given contravariant vector. As we will now explain the above formula has the closest connection to the law of the change of angles between vectors which are parallel transported.

The angle ω between two covariant vectors ξ^i and η^i is defined by the condition:

$$\cos \omega = \frac{\Delta(\xi^i, \eta^i)}{\sqrt{l(\xi^i)}\sqrt{l(\eta^i)}}, \qquad (17)$$

where $\Delta(\xi^i, \eta^i)$ is determined from the equation:

$$\Delta(\xi^i, \eta^i) = g_{ik}\, \xi^i\, \eta^k. \qquad (18)$$

The angle Ω between two contravariant vectors ξ_i and η_i is defined by the condition:

$$\cos \Omega = \frac{\nabla(\xi_i, \eta_i)}{\sqrt{L(\xi_i)}\sqrt{L(\eta_i)}}, \qquad (19)$$

where $\nabla(\xi_i, \eta_i)$ is determined by the equation:[5]

$$\nabla(\xi_i, \eta_i) = g^{ik}\, \xi_i\, \eta_k. \qquad (20)$$

Here we will not discuss the internal connection between (15) and (17); this connection is clarified on the basis of considerations, which can be found, for example, in Bianchi, *Lezioni di Geometria Differentiale*, t. 1, p. 332.

2. We will now explain how the norms of vectors and angles change when the vectors are parallel transported along a given curve. Simple

---

[5] We will not discuss here the additional conditions which make the values of ω and Ω, obtained from (17) and (19) fully defined.

calculations give the following equations:

$$\frac{d\Delta}{dt} = A_{iks} \frac{dx_s}{dt} \xi^i \eta^k,$$
$$\frac{dl}{dt} = A_{iks} \frac{dx_s}{dt} \xi^i \xi^k,$$
$$\frac{d\nabla}{dt} = B_s^{ik} \frac{dx_s}{dt} \xi_i \eta_k,$$
$$\frac{dL}{dt} = B_s^{ik} \frac{dx_s}{dt} \xi_i \xi_k,$$
(21)

where the tensors A_{iks} and B_s^{ik} are determined by the formulas:

$$A_{iks} = \frac{\partial g_{ik}}{\partial x_s} - g_{\alpha k} \Gamma_{is}^{\alpha} - g_{i\alpha} \Gamma_{ks}^{\alpha}, \qquad (22)$$

$$B_s^{ik} = \frac{\partial g^{ik}}{\partial x_s} + g^{\alpha k} G_{\alpha s}^i + g^{i\alpha} G_{\alpha s}^k \qquad (23)$$

In other words, A_{iks} is the tensorial derivative of g_{ik} formed with the help of the tensorial parameters $\Gamma_{\lambda\mu}^i$, and B_s^{ik} is the tensorial derivative of g^{ik} formed with the help of the tensorial parameters $G_{\lambda\mu}^i$.

Using the previous equations we find:

$$\frac{d\cos\omega}{dt} = \frac{1}{\sqrt{l_1}\sqrt{l_2}} \left(A_{iks} \xi^i \eta^k - \frac{1}{2}\frac{\Delta}{l_1} A_{pqs} \xi^p \xi^q - \frac{1}{2}\frac{\Delta}{l_2} A_{pqs} \eta^p \eta^q \right) \frac{dx_s}{dt},$$
$$\frac{d\cos\Omega}{dt} = \frac{1}{\sqrt{L_1}\sqrt{L_2}} \left(B_s^{ik} \xi_i \eta_k - \frac{1}{2}\frac{\nabla}{L_1} B_s^{pq} \xi_p \xi_q - \frac{1}{2}\frac{\nabla}{L_2} B_s^{pq} \eta_p \eta_q \right) \frac{dx_s}{dt},$$
(24)

where for brevity we denoted:

$$l_1 = l(\xi^i), \ l_2 = l(\eta^i), \ L_1 = L(\xi_i), \ L_2 = L(\eta_i),$$

$$\Delta = \Delta(\xi^i, \eta^i), \ \nabla = \nabla(\xi_i, \eta_i),$$

and ξ^i, η^i and ξ_i, η_i are not connected with any relations to each other.

3. The space, in which the angles between two vectors, both covariant and contravariant, do not change when these vectors are parallel transported, will be called *Weyl's space*. It is possible to consider

separately spaces, having this property only for covariant or only for contravariant vectors, but we will not do it here due to the lack of space.

The following proposition will state the main property of Weyl's space.

Theorem. *In order that a space be Weyl's space, it is necessary and sufficient for the ratios $\frac{A_{iks}}{g_{ik}}$ and $\frac{B_s^{ik}}{g^{ik}}$ be independent of the indices i and k.*

To prove the necessary condition of the theorem, let us consider the parallel transport of orthogonal covariant vectors, i.e. such vectors for which $\cos\omega = 0$ or $\Delta(\xi^i, \eta^i) = 0$. For Weyl's space we have $\frac{d\Delta}{dt} = 0$, or, using formula (24) and $\Delta = 0$, we find the following equation:

$$A_{iks} \frac{dx_s}{dt} \xi^i \eta^k = 0,$$

providing that ξ^i, η^i satisfy the relation:

$$g_{ik} \xi^i \eta^k = 0.$$

It follows from these equations that:

$$\frac{A_{iks}}{g_{ik}} \cdot \frac{dx_s}{dt} = \frac{A_{lms}}{g_{lm}} \cdot \frac{dx_s}{dt}$$

which equation holds for all $\frac{dx_s}{dt}$ and proves the necessary condition of the theorem for covariant vectors.

In exactly the same way we will prove the necessary condition of the theorem for contravariant vectors as well.

The sufficient condition of the theorem is verified directly by using formulas (24).

We will denote the expressions $\frac{A_{iks}}{g_{ik}}$ and $\frac{B_s^{ik}}{g^{ik}}$ by $-\varphi_s$ and $-f_s$, respectively. Then we will have:

$$A_{iks} = -\varphi_s g_{ik}, \quad B_s^{ik} = -f_s g^{ik}, \tag{25}$$

where φ_s and f_s will be contravariant vectors.

Using formulas (21) and (25) we find:

$$\frac{dl}{dt} = -\varphi_s \frac{dx_s}{dt} l, \quad \frac{dL}{dt} = -f_s \frac{dx_s}{dt} L. \tag{26}$$

In such a way φ_s and f_s will be contravariant vectors characterizing the change of the norm of vectors as long as these vectors are parallel transported along some curve. We will call these vectors *first and second scale vectors*. The connection between them holds for vectorially perfect spaces, which will be discussed below.

4. The tensors φ_{ik} and f_{ik} are defined by the equations:

$$\varphi_{ik} = \frac{\partial \varphi_i}{\partial x_k} - \frac{\partial \varphi_k}{\partial x_i},$$
$$f_{ik} = \frac{\partial f_i}{\partial x_k} - \frac{\partial f_k}{\partial x_i} \tag{27}$$

We will call them the *first and second metric curvature of space*. It is easy to see that the necessary and sufficient condition for the independence of the change of the norm of a covariant or contravariant vector on the path, along which a vector is parallel transported, is the zero value of the first or second metric curvature of the space, respectively.

We will say that we *change the scale*, if all g_{ik} are multiplied by $\lambda = \lambda(x_1, x_2, \ldots, x_n)$. In this case the norm of a covariant vector will be multiplied by λ, and the norm of a contravariant vector will be multiplied by $\frac{1}{\lambda}$, because all g^{ik} will acquire the quantity $\frac{1}{\lambda}$. The first scale vector φ_s turns into the vector

$$\varphi_s - \frac{\partial \log \lambda}{\partial x_s},$$

and the second scale vector f_s goes into the vector

$$f_s + \frac{\partial \log \lambda}{\partial x_s}.$$

As far as the first and second metric curvature are concerned, they do not change.

We will denote the expression, to which a quantity A changes, when we change the scale, by putting the sign $\sim$ above the letter representing the quantity (i.e. $\tilde{A}$). The quantities which do not change when the scale is changed, we will call (following Weyl's terminology) *scale invariants*, leaving the notion *coordinate invariant* for invariants in the usual sense.

Because further on we will have to construct scale invariants, we need to clarify how different expressions change when the scale is changed. It is easy to compose the following formulas:

$$\tilde{g}_{ik} = \lambda\, g_{ik}, \quad \tilde{g}^{ik} = \frac{1}{\lambda}\, g^{ik}, \quad \tilde{g} = \lambda^n\, g, \tag{28}$$

$$\tilde{R} = \frac{1}{\lambda} R + \frac{n-1}{\lambda} D\psi + \frac{(n-1)(n-2)}{4\lambda}\Delta\psi,$$

where $D\psi$ and $\Delta\psi$ are defined by the equations:

$$D\psi = \frac{1}{\sqrt{g}}\frac{\partial}{\partial x_\beta}\left(\sqrt{g}\, g^{\alpha\beta}\frac{\partial\psi}{\partial x_\alpha}\right),$$

$$\Delta\psi = g^{\alpha\beta}\frac{\partial\psi}{\partial x_\alpha}\frac{\partial\psi}{\partial x_\beta}, \tag{29}$$

$$\psi = \log\lambda,$$

and R is the scalar curvature, formed for the tensorial parameters

$$\Gamma^i_{\lambda\mu} = \begin{Bmatrix}\lambda\mu\\i\end{Bmatrix} : R = g^{ik}\, g^{\alpha\beta}\,(i\,\alpha,\ \beta\,k).$$

It is not difficult to see that the following equations hold:

$$\tilde{\varphi}_i = \varphi_i - \frac{\partial\psi}{\partial x_i}, \quad \tilde{f}_i = f_i + \frac{\partial\psi}{\partial x_i},$$

$$\tilde{\varphi}_{ik} = \varphi_{ik}, \quad \tilde{f}_{ik} = f_{ik}. \tag{30}$$

§4

1. Let us now define the tensorial parameters for Weyl's spaces. We start with the covariant tensorial parameters $\Gamma^i_{\lambda\mu}$; they should be defined from equation (25), i.e. from the equation:

$$\frac{\partial g_{ik}}{\partial x_s} + \varphi_s\, g_{ik} = g_{\alpha k}\,\Gamma^\alpha_{is} + g_{\alpha i}\,\Gamma^\alpha_{ks} \tag{31}$$

If $\Gamma^i_{\lambda\mu}$ are symmetric parameters, then from equation (31) Weyl finds the following expressions for $\Gamma^i_{\lambda\mu}$:

$$\Gamma^i_{\lambda\mu} = \Lambda^i_{\lambda\mu} = \begin{Bmatrix}\lambda\mu\\i\end{Bmatrix} + \frac{1}{2}\varphi_\lambda\,\delta^i_\mu + \frac{1}{2}\varphi_\mu\,\delta^i_\lambda - \frac{1}{2}g_{\lambda\mu}\,\varphi^i, \tag{32}$$

where φ^i is defined by:
$$\varphi^i = g^{\sigma i} \varphi_\sigma. \tag{33}$$

We will show that the general solution of equation (31) is given by the formula:
$$\Gamma^i_{\lambda\mu} = \Lambda^i_{\lambda\mu} + a^i_{\lambda\mu} - g_{\alpha\lambda} g^{\sigma i} a^\alpha_{\sigma\mu} - g_{\alpha\mu} g^{\sigma i} a^\alpha_{\sigma\lambda} \tag{34}$$

where the $a^i_{\lambda\mu}$ is an arbitrary tensor which is skew-symmetric in its low indices. It follows from formula (34) that in the case of symmetry of the tensorial parameters we have $a^i_{\lambda\mu} = a^i_{\mu\lambda}$, and therefore the arbitrary tensor becomes zero: $a^i_{\lambda\mu} = 0$.

A direct calculation can show that $\Gamma^i_{\lambda\mu}$ defined by formula (34) satisfies the relation (31) if it is taken into account that $a^i_{\lambda\mu} = -a^i_{\mu\lambda}$. We will show that each solution of equation (31) can be represented in the form (34). We easily find from (31):

$$\begin{bmatrix} \lambda\mu \\ i \end{bmatrix} + \frac{1}{2}\varphi_\mu g_{\lambda i} + \frac{1}{2}\varphi_\lambda g_{\mu i} - \frac{1}{2}\varphi_i g_{\lambda\mu}$$
$$= g_{\alpha i} \Gamma^\alpha_{\lambda\mu} - \frac{1}{2} g_{\alpha i} \gamma^\alpha_{\lambda\mu} - \frac{1}{2} g_{\alpha\lambda} \gamma^\alpha_{\mu i} - \frac{1}{2} g_{\alpha\mu} \gamma^\alpha_{\lambda i}$$

where $\gamma^i_{\lambda\mu} = \Gamma^i_{\lambda\mu} - \Gamma^i_{\mu\lambda}$.

Replacing i by σ in this equation, multiplying its two sides with $g^{i\sigma}$, and summing over σ from 1 to n, we obtain:

$$\Gamma^i_{\lambda\mu} = \Lambda^i_{\lambda\mu} + \frac{1}{2}\gamma^i_{\lambda\mu} - \frac{1}{2}g_{\alpha\lambda} g^{\sigma i} \gamma^\alpha_{\sigma\mu} - \frac{1}{2}g_{\alpha\mu} g^{\sigma i} \gamma^\alpha_{\sigma\lambda}$$

or, put differently, equation (34), in which $a^i_{\lambda\mu}$ is replaced by the tensor $\frac{1}{2}\gamma^i_{\lambda\mu}$, by its own definition is skew-symmetric.

It is not difficult to conclude from formula (34) that the covariant tensorial parameters of Weyl's space are determined by the metric tensor g_{ik}, the scale vector φ_i and the skew-symmetric in the low indices tensor $a^i_{\lambda\mu}$, i.e. the definition of the tensorial parameters requires the knowledge of

$$\frac{n(n+1)}{2} + n + \frac{n^2(n-1)}{2}$$

coordinate functions. If the tensorial parameters are symmetric, their definition requires the knowledge of

$$\frac{n(n+1)}{2} + n$$

coordinate functions.

2. We will now determine the contravariant tensorial parameters of Weyl's space. According to formula (25) this can be dome by obtaining the solution of the equations:

$$\frac{\partial g^{ik}}{\partial x_s} + f_s\, g^{ik} = -g^{\alpha k} G^i_{\alpha s} - g^{\alpha i} G^k_{\alpha s} \tag{35}$$

We will show that the general solution of equation (35) is determined from the following formula:

$$G^i_{\lambda\mu} = \Gamma^i_{\lambda\mu} + \delta^i_\lambda \psi_\mu + g^{i\sigma} b_{\sigma\lambda\mu}, \tag{36}$$

where the contravariant vector ψ_μ is determined from the following formula:

$$\psi_\mu = -\frac{\varphi_\mu + f_\mu}{2},$$

and the tensor $b_{\sigma\lambda\mu}$ is skew-symmetric in its first two indices.

Multiplying both sides of (35) by $g_{kl}\, g_{im}$, summing over i and k from 1 to n and adding the obtained result to formula (31), where i and k are replaced with l and m, we find the following relation:

$$g_{im}\, g_{kl}\, \frac{\partial g^{ik}}{\partial x_s} + \frac{\partial g_{lm}}{\partial x_s} + (\varphi_s + f_s)\, g_{lm} = -g_{\alpha m}\, \iota^\alpha_{ls} - g_{\alpha l}\, \iota^\alpha_{ms},$$

where $\iota^i_{\lambda\mu} = G^i_{\lambda\mu} - \Gamma^i_{\lambda\mu}$.

Direct calculation gives:

$$g_{im}\, g_{kl}\, \frac{\partial g^{ik}}{\partial x_s} + \frac{\partial g_{lm}}{\partial x_s} = 0$$

and therefore the previous equation can be rewritten as:

$$2\psi_s\, g_{lm} = g_{\alpha m}\, \iota^\alpha_{ls} - g_{\alpha l}\, \iota^\alpha_{ms}.$$

Putting
$$\iota^i_{\lambda\mu} = \delta^i_\lambda \psi_\mu + \vartheta^i_{\lambda\mu}$$
we find:
$$g_{\alpha m} \vartheta^\alpha_{ls} + g_{\alpha l} \vartheta^\alpha_{ms} = 0.$$
Introducing the tensor:
$$g_{\alpha m} \vartheta^\alpha_{ls} = b_{mls}, \quad \vartheta^i_{ls} = g^{\sigma i} b_{\sigma ls}$$
we find from the previous equations that it is skew-symmetric in its first two indices, and since
$$\iota^i_{\lambda\mu} = \delta^i_\lambda \psi_\mu + g^{\sigma i} b_{\sigma\lambda\mu}$$
we arrive at the conclusion that each solution of equation (35) is represented in the form of formula (36).

We see by a direct simple calculation that $G^i_{\lambda\mu}$ determined by formula (36) will be a solution of equation (35).

In such a way, in the general case of Weyl's space, the definition of contravariant tensorial parameters requires, in addition to the quantities defining the covariant parameters, also the knowledge of a second scale vector f_i and the tensor $b_{\sigma\lambda\mu}$ which is skew-symmetric in its first two indices; therefore it is necessary additionally to specify

$$n + \frac{n^2(n-1)}{2}$$

coordinate functions.

3. We will now clarify what limitations the vectorial perfect space imposes on the introduced above vector and tensor defining the tensorial parameters in Weyl's space, i.e. the parallel transport of vectors.

We will prove the following theorem:

Theorem. *If Weyl's space is a vectorially perfect space, then the semi-sum of the first and the second scale vectors is a characteristic vector with an opposite sign:*

$$\omega_i = \psi_i = -\frac{\varphi_i + f_i}{2}$$

In accordance with the formulas of the preceding section we have:

$$\iota^i_{\lambda\mu} = \delta^i_\lambda \psi_\mu + g^{i\sigma} b_{\sigma\lambda\mu}$$

but by the condition of vectorial perfect space we will have:
$$\iota^i_{\lambda\mu} + \iota^i_{\mu\lambda} = \delta^i_\lambda \omega_\mu + \delta^i_\mu \omega_\lambda.$$

From here we easily find:
$$b_{k\lambda\mu} + b_{k\mu\lambda} = g_{\lambda k} \chi_\mu + g_{\mu k} \chi_\lambda,$$
$$b_{\lambda k\mu} + b_{\lambda\mu k} = g_{k\lambda} \chi_\mu + g_{\mu\lambda} \chi_k,$$
$$b_{\mu k\lambda} + b_{\mu\lambda k} = g_{k\mu} \chi_\lambda + g_{\lambda\mu} \chi_k,$$

where $\chi_i = \omega_i - \psi_i$.

Adding the obtained formulas and taking into account that $b_{\sigma\lambda\mu}$ is skew-symmetric in its first two indices, we find:
$$g_{k\lambda} \chi_\mu + g_{\mu k} \chi_\lambda + g_{\lambda\mu} \chi_k = 0.$$

Multiplying this equation by $g^{\mu s} g^{\lambda\sigma}$, summing over μ and λ from 1 to n, and regarding s, σ as different from k we find:
$$\chi_k = 0,$$

and so on.

<u>Theorem</u>. *If the vectorially perfect Weyl's space has symmetric covariant parameters, then it is a principal vectorially perfect space.*

In other words, for such a space $\omega_i = \psi_i = 0$,
i.e., $\varphi_i = -f_i$.
Since $\Gamma^i_{\lambda\mu} = \Gamma^i_{\mu\lambda}$ and the space is vectorially perfect, then $\iota^i_{\lambda\mu} = \iota^i_{\mu\lambda}$ and therefore:
$$\iota^i_{\lambda\mu} = \frac{1}{2} \delta^i_\mu \omega_\lambda + \frac{1}{2} \delta^i_\lambda \omega_\mu.$$

Determining $\vartheta^i_{\lambda\mu}$ and also b_{mls} from here, we obtain the following equation:
$$b_{mls} = \frac{1}{2} g_{sm} \omega_l - \frac{1}{2} g_{lm} \omega_s.$$

Using the fact that b_{mls} is skew-symmetric in the first two indices we easily prove the theorem.

It is possible to assume that for Weyl's space in the case of non-symmetric tensorial parameters, the tensor of third rank $b_{\sigma\lambda\mu}$ is a very special case; but a closer look at such an assumption shows that it is not correct and for this reason we will not discuss it here.

So, if Weyl's space is vectorially perfect, its tensorial parameters are defined with the fundamental metric tensor, and with the first and second scale vectors, and with the tensor $a^i_{\lambda\mu}$, which is skew-symmetric in the low indices. In other words, for defining Weyl's vectorially perfect space, we need to know

$$\frac{n(n+1)}{2} + 2n + \frac{n^2(n-1)}{2}$$

coordinate functions. Later we will consider only such vectorially perfect Weyl's spaces, for which the tensor $a^i_{\lambda\mu}$ is expressed through scale vectors. That is, we will suppose that a space is defined by a metric tensor and two scale vectors.

§5

1. From a metric tensor and from the two scale vectors we can form a number coordinate invariants. From the components of the metric tensor, its first and second derivatives, it is easy to form the scalar curvature in the sense of Riemann R. From the scale vectors one can construct the following invariants:

$$\varphi = g^{ik}\varphi_i\varphi_k, \quad f = g^{ik}f_i f_k, \quad \mu = g^{ik}\varphi_i f_k. \tag{37}$$

Constructing, with the help of Christoffel's brackets, tensorial derivatives of the vectors φ_i and f_i and denoting them by φ'_i and f'_i, we find the following equations:

$$\varphi'_{ik} = \frac{\partial \varphi_i}{\partial x_k} - \varphi_\sigma \left\{ \begin{matrix} ik \\ \sigma \end{matrix} \right\}, \quad f'_{ik} = \frac{\partial f_i}{\partial x_k} - f_\sigma \left\{ \begin{matrix} ik \\ \sigma \end{matrix} \right\}.$$

Using these equations and and also the metric curvatures φ_{ik} and f_{ik} we can construct the following invariants, containing scale vectors, their first derivatives, the metric tensor and its first derivatives:

$$\Phi = g^{ik}\varphi'_{ik} = \frac{1}{\sqrt{g}}\frac{\partial(\sqrt{g}\, g^{\sigma k}\varphi_\sigma)}{\partial x_k},$$

$$\mathcal{F} = g^{ik}f'_{ik} = \frac{1}{\sqrt{g}}\frac{\partial(\sqrt{g}\, g^{\sigma k}f_\sigma)}{\partial x_k} \tag{38}$$

$$S_1 = \varphi_{ik}\,\varphi^{ik} = g^{i\alpha}\,g^{k\beta}\,\varphi_{\alpha\beta}\,\varphi_{ik},$$
$$S_2 = f_{ik}\,f^{ik} = g^{i\alpha}\,g^{k\beta}\,f_{\alpha\beta}\,f_{ik}, \qquad (39)$$
$$S_3 = \varphi_{ik}\,f^{ik} = f_{ik}\,\varphi^{ik} = g^{i\alpha}\,g^{k\beta}\,\varphi_{\alpha\beta}\,f_{ik}.$$

In this way nine coordinate invariants are formed. A number of scale invariants can be constructed from them. We will now form these scale invariants.

2. If the change of the scale by multiplying the metric tensor by a coordinate function λ affects a given coordinate invariant by multiplying it by λ^e, i.e. if we have the relation

$$\tilde{A} = \lambda^e\,A,$$

then we will call the coordinate invariant A *relative scale invariant of order e*.

It is easy to see that the coordinate invariants S_i, defined by formulas (39) are relative scale invariants of order -2.

Using the coordinate invariants (37), (38) and also the invariant R it is easy to form linear combinations, which will be relative scale invariants. Carrying out simple calculations we have:

$$\tilde{\varphi} = \frac{1}{\lambda}\,(\varphi + \Delta\psi - 2\,Z_1\,\psi),$$
$$\tilde{\Phi} = \frac{1}{\lambda}\left(\Phi - D\psi - \frac{n-2}{2}\,\Delta\psi + \frac{n-2}{2}\,Z_1\,\psi\right),$$
$$\tilde{f} = \frac{1}{\lambda}\,(f + \Delta\psi + 2\,Z_2\,\psi), \qquad (40)$$
$$\tilde{\mathcal{F}} = \frac{1}{\lambda}\left(\mathcal{F} + D\psi - \frac{n-2}{2}\,\Delta\psi - \frac{n-2}{2}\,Z_2\,\psi\right),$$
$$\tilde{\mu} = \frac{1}{\lambda}\,(\mu - \Delta\psi - Z_2\,\psi + Z_1\,\psi),$$

where the operations Z_1 and Z_2 are defined by the equations:

$$Z_1\,\psi = \varphi^\alpha\,\frac{\partial\psi}{\partial x_\alpha} = g^{\sigma\alpha}\,\varphi_\sigma\,\frac{\partial\psi}{\partial x_\alpha},$$
$$Z_2\,\psi = f^\alpha\,\frac{\partial\psi}{\partial x_\alpha} = g^{\sigma\alpha}\,f_\sigma\,\frac{\partial\psi}{\partial x_\alpha}, \qquad (41)$$

and we already introduced the rest of the definitions at the end of §3. Using formulas (28) and (40) we form the following two relative scale invariants of order -1:

$$\Lambda_1 = R + (n-1)\Phi + \frac{(n-1)(n-2)}{4}\varphi, \qquad (42)$$

$$\Lambda_2 = \varphi + f + 2\mu.$$

The first of this invariants is nothing else but the invariant considered by Weyl, which he denoted by F (curvature of the tensorial parameters of space discussed by Weyl).[6] The second invariant of space in Weyl's work becomes identically zero.

From the two relative scale invariants of order -1 and the three relative scale invariants of order -2, obtained by us, it is not difficult to form an absolute scale invariant, or, generally, a relative scale invariant of order e according to the following formula:

$$\Lambda = \Lambda_1^{-e}\,\mathcal{O}\,(\Lambda_1,\,\Lambda_2,\,S_1^{\frac{1}{2}},\,S_2^{\frac{1}{2}},\,S_3^{\frac{1}{2}}), \qquad (43)$$

where $\mathcal{O}$ is a zero dimension homogeneous function of its arguments. Obviously, an integral invariant is the expression:

$$\mathcal{J} = \int \mathfrak{M}\,dV = \int \Lambda_1^{\frac{n}{2}},\,\mathcal{O}\,(\Lambda_1,\,\Lambda_2,\,S_1^{\frac{1}{2}},\,S_2^{\frac{1}{2}},\,S_3^{\frac{1}{2}})\,\sqrt{g}\,dV,$$

where $dV = dx_1\,dx_2,\ldots,dx_n$

Weyl uses in his work a special form of the integral function $\mathfrak{M}$ which can be written in our notations as:

$$\mathfrak{M} = \Lambda_1^2\,\sqrt{g}\left(a + b\left(\frac{S^{\frac{1}{2}}}{\Lambda_1}\right)^2\right),$$

where a and b are constant quantities.

3. In conclusion I will take the liberty, althought this is not related to the subject of the present Note, to touch on the question of the physical significance of the developed above geometrical considerations. From the zero value of the variation of the integral invariant $\mathcal{J}$, Weyl obtains, on the one hand, Einstein's equations, and on the

---

[6]The sign of the curvature in Weyl's equation is opposit to the one used here.

other hand Maxwell's equations, in the form in which they are written in Mie's theory. In this way Weyl identifies the scale vector with the four-dimensional electromagnetic potential. By using general considerations regarding the properties of the integral invariant, Weyl obtains the general form of the equations of electrodynamics, and by deriving a special form of the world function $\mathfrak{M}$ he arrives at Maxwell's equations and Mie's theory. As in the developed by us geometry there exist two scale vectors, it is completely natural to think of the possibility to identify them with a four-dimensional magnetic potential and a four-dimensional vector of the current (Viererstrom). It is easy to obtain, using infinitesimal scale changes, a number of indications regarding the character of Maxwell's equations and the additional relations, which from our point of view can replace Mie's theory. A significantly more difficult task is the choice of such a special form of the world function $\mathfrak{M}$, which could yield the additional relations, mentioned above, without the contradictions in the realm of the electron theory to which, unfortunately, leads Mie's theory.

In the present Note I leave the question of the possibility of finding the world function $\mathfrak{M}$ with the just outlined properties open.

Petrograd,   A. Friedmann,
15 April 1922   Professor of Mechanics of the
                                                      Petrograd Polytechnic Institute

Image taken from Friedmann's manuscript typed in Russian and preserved in the Ehrenfest archive.

MAIN DATES IN FRIEDMANN'S LIFE AND WORK[1]

Alexander Alexandrovich Friedmann was born on June 4 (16), 1888 in St. Petersburg.

1897-1906 Student at the 2nd St. Petersburg (Alexander I) Gymnasium

1905- First Scientific study written (published in 1906)

1905-06 Takes part in the students' movement of St. Petersburg secondary schools; member of the Central Committee of this organization

1906-10 Student at the Faculty of Physics and Mathematics of St. Petersburg University

1910-13 Retained at the University to prepare for professor's work

1910-14 Teaching practical classes in mathematics; lecturer at the Institute of Railway Engineering

1912-14 Lecturer at the Institute of Mining

1913 Passes Master's examinations and becomes a Master's degree student at St. Petersburg University

1913-14 Staff worker at the Nicholas Aerological Observatory in Pavlovsk (a suburb of St. Petersburg)

---

[1]This list is taken from E.A. Tropp, V.Ya. Frenkel, A.D. Chernin, *Alexander A. Friedmann: The Man who Made the Universe Expand.* Translated by A. Dron and M. Burov (Cambridge University Press, Cambridge 1993) pp. 256-257.

1914 Sent to Leipzig (Germany), scientific research work under V. Bjerknes

1914-16 Takes part in World War I; volunteers to join the Army, serving in aviation units of the northern and southern fronts

1916-17 In charge of the Central Aeronautical Service (CAS) of the front; instructor at the Kiev school of observer-pilots

1917-18 Employee, and later Acting Director, of the Moscow "Aviapribor" plant under the Air Force Administration

1918-20 Professor and the Chair of Mechanics at Perm State University

1920-24 Research worker in the Atomic Commission at the State Optical Institute

1920-24 Teaching mathematics and mechanics at the Faculty of Physics and Mechanics of Petrograd University

1920-25 Professor of the Faculty of Physics and Mathematics of the Petrograd Polytechnical Institute

1920-25 Instructor, and from 1921 Professor, in the Department of Applied Aerodynamics in the Faculty of Air Communications at the Petrograd Institute of Railway Engineering

1920-25 Junior scientific assistant, senior supervisor of studies at the *Naval Academy*

1920-25 Senior physicist, head of the mathematical bureau, learned secretary and from February 1925 Director of the Main Physical (later Geophysical) Observatory

1922 Writes and publishes the article "On the curvature of space"

1922 Defends his Master's dissertation; the corresponding book *The Hydromechanics of a Compressible Fluid* is published

1923 The book *The World as Space and Time* is published

1923-25 Editor of the journal *Geophysics and Meteorology*

1923 Scientific mission to Germany and Norway

1924 Attends the *International Congress for Applied Mechanics* in Delft (Netherlands), elected member of the standing committee for convening the Congress

1924 First volume of the book *Fundamentals of the Theory of Relativity* is published (written jointly with V. K. Frederiks)

1925 Editor-in-chief of the Journal *Klimat i Pogoda* (Climate and Weather)

1925 Edits the geophysics section of the great *Soviet Encyclopedia*
July 1925 Ascends to a record-breaking altitude of 7,400 m in a balloon and conducts investigations there (jointly with P. F. Fedoseyenko)

On September 16, 1925, Alexander Alexandrovich Friedmann died in Leningrad. Buried (on September 18) in the Smolensky cemetery.

Image Credits

Photograph after the title page: http://publ.lib.ru/ARCHIVES/F/FRIDMAN_Aleksandr_Aleksandrovich/
p. 2: Image taken from Friedmann's Biography[2], p. 55.
p. 3: Image taken from: http://nuclphys.sinp.msu.ru/persons/images/fridman_aleksandr.jpg
p. 4: Image taken from Friedmann's Biography, p. 98
p. 4: Image taken from Friedmann's Biography, p. 208
p. 5: Image taken from Friedmann's Biography, p. 118
p. 6 : http://books.e-heritage.ru/book/10079957
p. 8: Image taken from Friedmann's Biography, p. 160

---

[2]E.A. Tropp, V.Ya. Frenkel, A.D. Chernin, *Alexander A. Friedmann: The Man who Made the Universe Expand*. Translated by A. Dron and M. Burov (Cambridge University Press, Cambridge 1993) pp. 256-257.

CPSIA information can be obtained
at www.ICGtesting.com
Printed in the USA
LVOW12s1531100416
482975LV00026B/649/P